Experimental Results for Phase Equilibria and Pure Component Properties

John R. Cunningham and Dennis K. Jones, editors

R. D. Chircio
T. E. Daubert
I. A. Hossenlopp
T. Jianhua
A. D. Leu
W. Lihua
A. Nguyen
R. H. Powell
D. B. Robinson
D. J. Rosenthal
R. L. Rowley
D. Shukla
N. K. Smith
W. V. Steele
A. S. Teja
W. V. Wilding
G. M. Wilson
L. C. Wilson
W. Zhaoli
C. Zhongxiu
D. Zudkevitch

AIChE Staff

Maura Mullen, Publications Production Manager

DIPPR Data Series

Number 1

1991

Published by the American Institute of Chemical Engineers
345 East 47th Street, New York, N.Y. 10017

Copyright 1991

American Institute of Chemical Engineers
345 East 47 Street, New York, N.Y. 10017

AIChE shall not be responsible for statements or opinions advanced in papers or printed in its publications.

Library of Congress Cataloging-in-Publication Data

Experimental results for phase equilibria and pure component properties/John R. Cunningham and Dennis K. Jones, editors.
    p. cm. -- (Data series: no. 1, 1991)
   Includes index.
   ISBN 0-8169-0563-0
   1. Phase rule and equilibrium. 2. Vapor pressure. I. Cunningham, John R., 1949-   . II. Jones, Dennis K. III. Series: Data series (Design Institute for Physical Property Data (U.S.)); 1991, no. 1.
QD503.E96   1991
660′.2963—dc20

91-39155
CIP

FOREWORD

This is the first volume of the AIChE DIPPR Data Series. The Data Series will present results from projects sponsored by the AIChE Design Institute for Physical Property Data (DIPPR). This volume contains data from each of the current experimental projects.

In the Phase Equilibrium section the experimental data for twenty-three binary systems are reported. Fourteen of these systems were funded by the sponsoring companies while the remaining nine were funded through the DIPPR 805 National Science Foundation Grant. The availability of these data will allow extension and prediction of correlations for predicting such equilibria.

In the Pure Component Properties section, liquid vapor pressure measurement results for 13 compounds obtained in DIPPR Project 821 are presented. Also reported are critical property measurements on ten compounds from DIPPR Project 851 and enthalpy of formation data for eight compounds measured in DIPPR Project 871.

This Data Series volume is the result of the efforts of the Steering Committee members of the four DIPPR projects. They have contributed in proposing areas for study, coordinating the research, and reviewing the data and correlations. The contributions of those listed on the following pages are greatly appreciated. We would also like to thank Maura Mullen of the AIChE Publications staff and Ted Selover, DIPPR Technical Director, for their valuable assistance in the preparation of this volume.

John R. Cunningham
Simulation Sciences Inc.
1051 W. Bastanchury Road
Fullerton, CA 92633

and

Dennis K. Jones
Eastman Chemical Company
P.O. Box 1972
Kingsport, TN 37662

CONTENTS

FOREWORD ii

PROJECT SPONSORS v

SUMMARY OF EXPERIMENTAL DATA REPORTED vii

I. Phase Equilibria

VAPOR-LIQUID EQUILIBRIUM IN SELECTED BINARY SYSTEMS OF INTEREST
   TO THE CHEMICAL INDUSTRY .... A. D. Leu and D. B. Robinson   1

VAPOR-LIQUID EQUILIBRIUM MEASUREMENTS ON EIGHT BINARY MIXTURES:
   DIPPR Projects 805(A)/89 and 805(E)/89 .... W. Vincent Wilding, Loren C. Wilson and Grant M. Wilson   6

ISOTHERMAL VAPOR-LIQUID EQUILIBRIUM OF PROPYLENE
   OXIDE-ACETOPHENONE SYSTEM .... Chen Zhongxiu, Tang Jianhua, Wang Lihua and Wu Zhaoli   24

VAPOR-LIQUID EQUILIBRIUM IN BINARIES OF SULFOLANE WITH
   ORTHO-XYLENE, OCTANE AND ETHYL-CYCLOHEXANE:
   Report on Subproject DIPPR 805NSF/88/B .... David Zudkevitch and Deepak Shukla   32

DIPPR Project 805 NSF (B)/89-
   VAPOR-LIQUID EQUILIBRIUM MEASUREMENTS ON MIXTURES IMPORTANT TO INDUSTRIAL DESIGN:
   PHENOL + DIMETHYLETHER, PHENOL + PROPYLAMINE, ETHYLFORMATE + DIMETHYLETHER, ETHYLBENZENE
   + PROPYLAMINE, METHYLAMINE + DIMETHYLETHER, AND ANILINE + DIMETHYLETHER
   .... Richard L. Rowley and Roger H. Powell   47

II. Pure Component Properties

VAPOR PRESSURE OF 13 PURE INDUSTRIAL CHEMICALS .... Thomas E. Daubert   80

THE CRITICAL PRESSURES AND TEMPERATURES OF TEN SUBSTANCES
   USING A LOW RESIDENCE TIME FLOW APPARATUS .... Amyn S. Teja and Daniel J. Rosenthal   96

DIPPR PROJECT 871: DETERMINATION OF IDEAL-GAS ENTHALPIES OF FORMATION FOR KEY COMPOUNDS:
   THE 1989 PROJECT RESULTS
   .... W. V. Steele, R. D. Chirico, A. Nguyen, I. A. Hossenlopp and N. K. Smith   101

INDEX 135

Members of DIPPR Steering Committee for Project 805
Experimental Data on Mixtures Project
(1988-1989)

N. Barona	D. K. Jones
G. Bentzen	T. T. Shih
C. H. Chan	R. Srivastava
T. Chang	G. A. Sweaney
W. M. Clarke	G. Thomson
J. R. Cunningham	K. W. Won
T. Fjeldberg	M. Takeuchi
Y. Izawa	

T. Thomas Shih, Chairman Steering Committee, 1988
John R. Cunningham, Chairman Steering Committee, 1989

DIPPR Project 805 Sponsors
(1988-1989)

Allied-Signal Corporation	Norsk Hydro, A. S.
ARCO Chemical Company	Olin Corporation
The Dow Chemical Company	Phillips Petroleum Company
E. I. Dupont de Nemours & Company, Inc.	Simulation Sciences Inc.
Eastman Kodak Company	Toyo Engineering Corporation
Ethyl Corporation	Union Carbide Corporation (1984)
Mitsui Engineering and Shipbuilding Company, Ltd.	

Members of DIPPR Steering Committee for Project 821
Pure Component Liquid Vapor Pressure Measurements Project
(1988-1989)

Gordon Bentzen	Dennis K. Jones
C. H. Chan	Navin C. Patel
James Conover	George H. Thomson
Beryl Edmonds	

Dennis K. Jones, Chairman, Steering Committee, 1988-1989

DIPPR Project 821 Sponsors
(1988-1989)

Allied-Signal Inc.	Pennwalt Corporation
Eastman Kodak Company	Phillips Petroleum Company
The Institution of Chemical Engineers	Union Carbide Chemicals & Plastics Company Inc.
Olin Corporation	

Members of DIPPR Steering Committee for Project 851
Critical Properties of Pure Components Project
(1989)

Monica E. Baltatu	Sanjay P. Godbole
C. H. Chan	Dennis K. Jones
Te Chang	Navin C. Patel
J. David Chase	Ted B. Selover
Al L. Coignet	George H. Thomson
Beryl Edmonds	Costa Tsonopoulos
Bruce A. Feay	Kai W. Young

Dennis K. Jones, Chairman Steering Committee, 1989

DIPPR Project 851 Sponsors
(1989)

Allied-Signal Inc.	Fluor Daniel, Inc.
Amoco Chemicals Company	Hoechst Celanese Corporation
ARCO Chemical Company	The Institution of Chemical Engineers
British Petroleum Co.	Phillips Petroleum Company
Dow Chemical Company	Phone-Poulenc, Inc.
Eastman Kodak Company	Union Carbide Chemicals & Plastics Company Inc.
Exxon Research & Engineering Company	

Members of DIPPR Steering Committee for Project 871
Pure Component Ideal Gas Heat of Formation Project
(1989)

Bruce A. Feay	George H. Thomson
Dennis K. Jones	Costa Tsonopoulos
Navin C. Patel	Kai W. Young
Ted. B. Selover	

Dennis K. Jones, Chairman, Steering Committee, 1989

DIPPR Project 871 Sponsors
(1989)

Amoco Chemicals Company	Phillips Petroleum Company
Eastman Kodak Company	Rhone-Poulenc, Inc.
Exxon Research & Engineering Company	Union Carbide Chemicals & Plastics Company Inc.

Summary of Experimental Data Reported

Phase Equilibria

Project/Year	Investigator	Systems
805(A)/88	A. D. Leu D. B. Robinson	Acetaldehyde + Acetone Acetaldehyde + Propylene Oxide 1,2,Butylene Oxide + Isobutylene Oxide 1,2,Butylene Oxide + Methyl Acetate
805 NSF(B)/88	D. Zudkevitch D. Shulka	Sulfolane + o-Xylene Sulfolane + n-Octane Sulfolane + Ethylcyclohexane
805 (A)/89 805 (E)/89	W. V. Wilding L. C. Wilson G. M. Wilson	Tetrahydrofuran + 2,3-Dihydrofuran 2-Methyltetrahydrofuran + Tetrahydrofuran Water + 2,5-Dihydrofuran Diethanolamine + Water 1,2-Propanediol + 1-methoxy-2-propanol Dipropylene glycol monomethyl ether + 1,2-Propandiol Acetophenone/Propene Dimethyl disulfide + Methanethiol
805 (D)/89	C. Zhongxiu T. Jianhua W. Lihua W. Zhaoli	Propylene Oxide + Acetophenone
805 NSF(B)/89	R. L. Rowley R. H. Powell	Phenol + Dimethylether Phenol + Propylamine Ethylformate + Dimethylether Ethylbenzene + Propylamine Methylamine + Dimethylether Aniline + Dimethylether

Liquid Vapor Pressure

Project/Year	Investigator	Compounds
821/1988-89	Thomas E. Daubert Gary Hutchison Tom Long Dan Blessner	sec-Butyl acetate 2-Methylbenzofuran Diacetone alcohol N-Aminoethyl ethanolamine 2-Phenylpropionaldehyde Methanesulfonyl chloride 1,2-Ethanedithiol t-Butyl acetate Benzylamine Methylacetoacetate n-Butylamine Ethyl-t-butyl ether Ethylthioethanol

Summary of Experimental Data Reported (continued)

Critical Properties

Project/Year	Investigator	Compounds
851/1989	Amyn S. Teja Daniel J. Rosenthal	Acetophenone 2-Butoxyethanol 2-Propoxyethanol Ethylene glycol Ethyl-3-ethoxy propionate 1-Methoxy 2-propyl acetate 2-(2-Propoxyethoxy) ethanol 2-(2-Butoxyethoxy) ethanol Monoethanolamine Valeric acid

Enthalpy of Formation

Project/Year	Investigator	Compounds
871/1989	W. V. Steele R. D. Chirico A. Nguyen I. A. Hossenlopp N. K. Smith	Butan-2-ol Tetradecan-1-ol Hexan-1,6-diol Methacrylamide Benzoyl formic acid Naphthalene-2,6-dicarboxylic acid Naphthalene-2,6-dicarboxylic acid dimethyl ester Tetraethylsilane

VAPOR-LIQUID EQUILIBRIUM IN SELECTED BINARY SYSTEMS OF INTEREST TO THE CHEMICAL INDUSTRY

A. D. Leu and D. B. Robinson ■ The University of Alberta,
Department of Chemical Engineering,
Edmonton, Alberta, Canada T6G 2G6

This paper presents experimental vapor-liquid equilibrium data on the acetaldehyde-acetone, acetaldehyde-propylene oxide, 1,2 butylene oxide-isobutylene oxide and 1,2 butylene oxide-methyl acetate binary systems. Measurements were made at 25° and 75°C for all systems. Pressures covered a range from a low of 25.2 kPa to a high of 540 kPa. The phenomenon of a double azeotrope was observed in the 1,2 butylene oxide-methyl acetate system.

The Design Institute for Physical Property Data of the American Institute of Chemical Engineers is interested in experimental vapor-liquid equilibrium data on selected binary systems of interest to the chemical process industry. The data obtained from the studies will be useful for improving the reliability of the design and hence the performance and efficiency of industrial plants, primarily those involved in petroleum or petrochemical processing. Another use for the data will be for improving the parameters available for use in equations of state or group contribution methods to predict the physical properties and equilibrium phase behavior of multicomponent systems containing the selected binary pairs.

SCOPE

The work covered by this investigation included experimental vapor-liquid equilibrium phase composition measurements on four binary systems at temperatures of 25° and 75°C. The systems were as follows:

Acetaldehyde	- Acetone
Acetaldehyde	- Propylene Oxide
1,2 Butylene Oxide	- Isobutylene Oxide
1,2 Butylene Oxide	- Methyl Acetate

University of Alberta
Edmonton, Alberta, Canada

In each case the equilibrium vapor and liquid compositions were measured from the vapor pressure of the less volatile component to the vapor pressure of the more volatile component or to the maximum (minimum) azeotropic pressure, whichever was higher (lower).

From 8 to 10 pressures were studied at each temperature, depending on the complexity of the phase diagram. Thus a complete set of P-x-y data were available for each binary at two temperatures.

EXPERIMENTAL

Equipment

A pistoned sapphire cell was used during the course of the study. This cell consisted of a fully transparent sapphire cylinder mounted firmly between two steel headers. The top header contained the necessary openings and valves for charging the components to the cell, for measuring the cell pressure and for removing samples for analysis. The bottom header accommodated a movable stainless steel piston which was driven manually by an externally mounted hand wheel. The body of the cell was 2.5 cm inside diameter with a length of 15.2 cm. The working volume of the cell was about 45 cm³.

The main cell and all the necessary lead lines and valves were mounted inside a controlled temperature air bath. The entire assembly including the cell and bath were rocked about a horizontal axis to achieve equilibrium of the cell contents.

This cell and its auxiliary equipment has been described in some detail by Huang et al.(_1_)

Measurements

Temperatures were measured using iron-constantan thermocouples which had been calibrated against a platinum resistance thermometer. Each temperature was read out on a digital voltmeter. They are believed known to within ±0.1°C.

Pressures were measured using various pressure transducers depending on the range and magnitude of the pressures encountered during the experiment. They were as follows:

A. Druck Pressure Transmitter, Model PTX 110/w, Druck Limited. Range 0-15 psia; Accuracy ±0.02 psia.

B. Statham Absolute Pressure Transducer, Model PA208TC-50-350, Statham Instruments, Inc. Range 0-50 psia; Accuracy ±0.1 psia.

C. Statham Absolute Pressure Transducer, Model PA208TC-100-350, Statham Instruments, Inc. Range 0-100 psia; Accuracy ±0.2 psia.

The specific pressure transducers used to measure pressures for each system were as follows:

System	Range, psi	Type
Acetaldehyde-Acetone	0-15; 0-50	A,B
Acetaldehyde-Propylene Oxide	0-15; 0-50	A,B
1,2 Butylene Oxide-Isobutylene Oxide	0-15; 0-100	A,D
1,2 Butylene Oxide-Methyl Acetate	0-15; 0-100	A,D

All analytical work was carried out on a Hewlett-Packard Model 7620A Gas Chromatograph coupled with an HP-3353 Data Acquisition System. A thermal conductivity detector was used for all four systems. Calibrations were made for each component and relative response factors were obtained from peak areas vs sample size. They were as follows:

System Component 1	Component 2	Response Factor Relative to Component 1
Acetaldehyde	Acetone	0.7504
Acetaldehyde	Propylene Oxide	0.7577
1,2 Butylene Oxide	Isobutylene Oxide	1.5913
1,2 Butylene Oxide	Methyl Acetate	1.0463

At least six samples of each phase were taken for analysis and the reported compositions are the result of averaging these six measurements. The repeatability of the analysis was generally within better than 3% of the relative standard deviation.

Materials

The purity and source of the materials used for the experimental work were as follows:

Chemical	Grade	Min. Purity Mole %
Acetaldehyde	Gold Label	99+
Acetone	HPLC	99.9
Propylene Oxide	Gold Label	99+
1,2,Butylene Oxide	Gold Label	99+
Methyl Acetate	Anhydrous	99+
Isobutylene Oxide		95

All chemicals, except isobutylene oxide, were obtained from Aldrich Chemical Co. The isobutylene oxide was obtained as a courtesy through ARCO Chemical Company.

Procedures

Prior to commencing an experimental run, the equipment was thoroughly cleaned and evacuated. A suitable amount of the less volatile component was then added to the cell, followed by a suitable quantity of the more volatile component. The contents of the cell were mixed by rocking the cell assembly and the cell was simultaneously brought to the desired temperature and pressure. When equilibrium had been established, the rocking motion was stopped and the sampling process commenced.

The gas phase was sampled by continuously drawing off a vapor stream through the sample valve under isobaric isothermal conditions where it was mixed with a stream of helium and circulated through the chromatographic valve. Several samples were taken for analysis at periodic intervals.

After the vapor phase had been analyzed, the remainder of the vapor phase plus a small interface portion of the liquid were removed from the cell. The liquid phase was then sampled and analyzed by using a similar procedure.

At the completion of each pressure point, a new set of conditions was established by adjusting pressure and/or adding more material. The equilibration and sampling sequence was then repeated.

EXPERIMENTAL RESULTS

The experimental data obtained during the study are presented in Tables 1 to 4 and are shown graphically in Figures 1 to 5. The additional figure is provided to provide better representation of the double azeotrope.

DISCUSSION

During the course of the work on the 1,2 butylene oxide - methyl acetate system, a very rare and unusual phenomenon was observed. Both a minimum and a maximum azeotrope occurred at approximately 15 mole % butylene oxide and the maximum at approximately 57 mole % butylene oxide.

To the best of our knowledge, only one other binary system exhibiting the double azeotrope phenomenon has been reported previously in the technical literature. This is for the system benzene - hexafluorobenzene reported by Gaw and Swinton(2). The occurrence of a double azeotrope in this system may arouse some theoretical interest in the results.

LITERATURE CITED

(1) Huang, S.S., Leu, A.D., Ng, H.-J., and Robinson, D.B., Fluid Phase Equilibria, 19, 21 (1985).

(2) Gaw, W.J. and Swinton, F.L., Nature, 212, 283 (1966).

Table 1. Equilibrium Phase Properties of the Acetaldehyde-Acetone System

Pressure kPa	Composition* Liquid	Vapor
	T = 25.0°C	
30.4	0.0	0.0
35.1	0.0522	0.1384
49.8	0.2234	0.4502
75.6	0.5114	0.7299
97.8	0.7598	0.8958
110	0.8813	0.9520
116	0.9655	0.9864
121	1.0	1.0
	T = 75.0°C	
183	0.0	0.0
207	0.0629	0.1398
249	0.1841	0.3517
323	0.3929	0.5966
430	0.7017	0.8407
480	0.8411	0.9234
528	0.9669	0.9846
540	1.0	1.0

* Mole Fraction: Acetaldehyde

Table 2. Equilibrium Phase Properties of the Acetaldehyde-Propylene Oxide

Pressure kPa	Composition* Liquid	Vapor
	T = 25.0°C	
71.7	0.0	0.0
75.5	00532	0.0881
83.5	0.2021	0.2866
95.1	0.4221	0.5253
107	0.6890	0.7605
113	0.8307	0.8701
119	0.9877	0.9898
120	1.0	1.0
	T = 75.0°C	
350	0.0	0.0
366	0.0842	0.1217
403	0.2864	0.3780
458	0.5807	0.6537
510	0.8540	0.8864
531	0.9690	0.9762
535	0.9909	0.9925
536	1.0	1.0

* Mole Fraction: Acetaldehyde

Table 3. Equilibrium Phase Properties of the 1,2 Butylene Oxide - Isobutylene Oxide System

Pressure kPa	Composition* Liquid	Vapor
	T = 25.0°C	
31.6	0.0	0.0
32.5	0.1372	0.1737
33.5	0.3479	0.4159
34.3	0.5056	0.5816
34.9	0.6474	0.7152
35.4	0.8191	0.8529
35.8	0.9327	0.9437
36.1	1.0	1.0
	T = 75.0°C	
142	0.0	0.0
145	0.0864	0.1108
154	0.2897	0.3437
161	0.4989	0.5520
168	0.7028	0.7459
171	0.8169	0.8375
174	0.9024	0.9131
177	1.0	1.0

* Mole Fraction: Isobutylene Oxide

Table 4. Equilibrium Phase Properties of the 1,2 Butylene Oxide - Methyl Acetate System

Pressure kPa	Composition* Liquid	Vapor
	T = 25.0°C	
28.6	0.0	0.0
26.3	0.0645	0.0413
25.2	0.1866	0.1977
31.4	0.2558	0.2693
43.4	0.4505	0.4703
43.9	0.6859	0.6453
31.4	0.9286	0.9098
27.3	0.9996	0.9969
27.0	1.0	1.0
	T = 75.0°C	
179	0.0	0.0
178	0.0983	0.0635
176	0.2818	0.1518
172	0.4901	0.2950
163	0.7186	0.5117
157	0.8677	0.7163
150	0.9438	0.8618
148	0.9791	0.9420
146	0.9946	0.9840
145	1.0	1.0

* Mole Fraction: Butylene Oxide

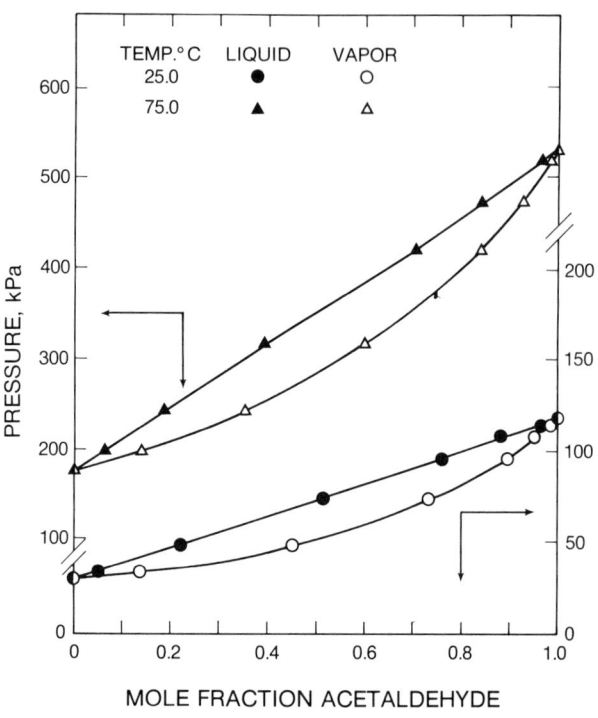

Figure 1. Equilibrium Phase Compositions For The Acetaldehyde-Acetone Binary System

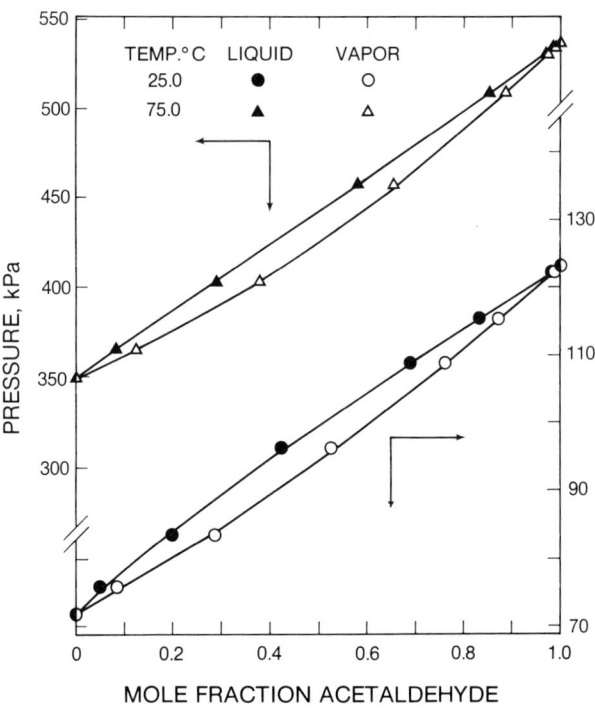

Figure 2. Equilibrium Phase Compositions For The Acetaldehyde-Propylene Oxide Binary System

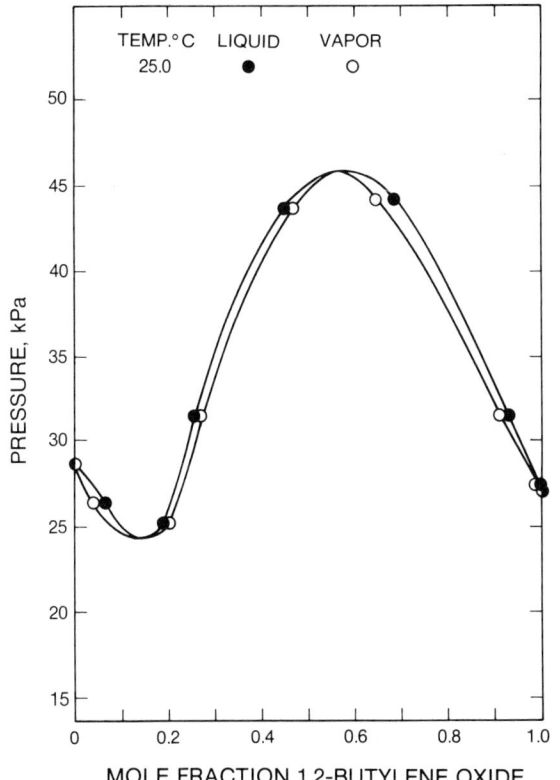

Figure 3. Equilibrium Phase Compositions For The 1,2 Butylene Oxide-Isobutylene Oxide Binary System

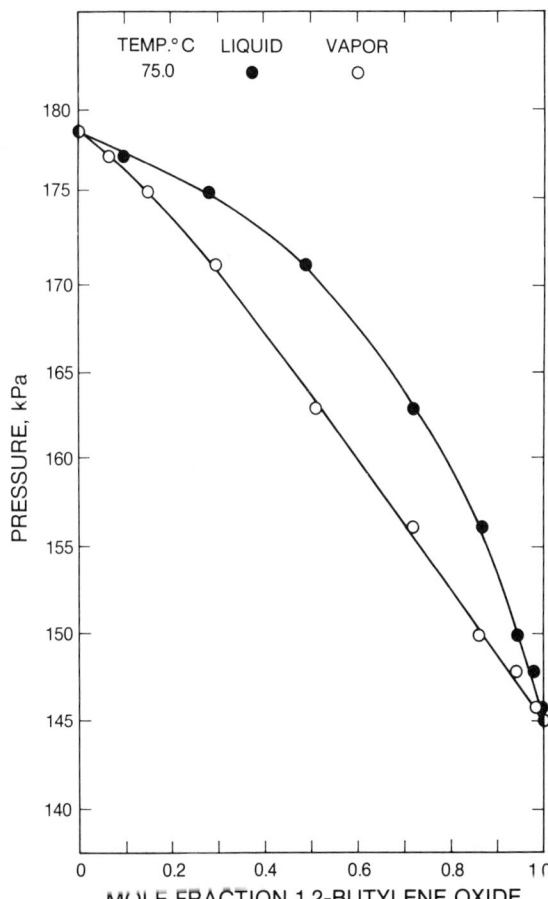

Figure 5. Equilibrium Phase Compositions For The 1,2 Butylene Oxide-Methyl Acetate Binary System At 75°C

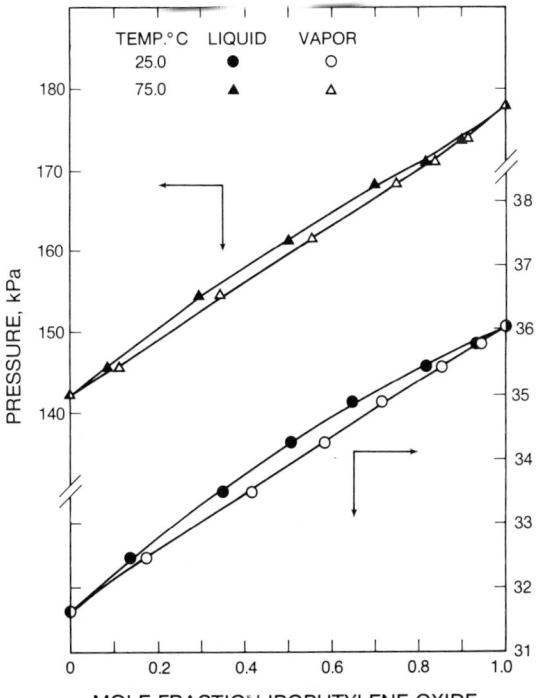

Figure 4. Equilibrium Phase Compositions For The 1,2 Butylene Oxide-Methyl Acetate Binary System At 25°C

VAPOR-LIQUID EQUILIBRIUM MEASUREMENTS ON EIGHT BINARY MIXTURES: DIPPR PROJECTS 805(A)/89 AND 805(E)/89

W. Vincent Wilding, Loren C. Wilson and Grant M. Wilson ■ Wiltec Research Co., Inc.,
488 South 500 West,
Provo, Utah 84601

Vapor-liquid equilibrium measurements have been made on eight binary mixtures at two temperatures each by the PTx method. Liquid-liquid equilibrium measurements on one of the systems were made at the lower temperature due to immiscibility. Equilibrium vapor and liquid phase compositions were derived from the data using the Soave equation of state to represent the vapor phase and the Wilson or the NRTL activity coefficient model to represent the liquid phase.

The Design Institute for Physical Property Data (DIPPR) of the American Institute of Chemical Engineers (AIChE) has sponsored this research to measure vapor-liquid equilibria on eight binary systems at two temperatures each. The systems studied and the temperatures at which they were studied are given in the table below.

System	Isotherms, °C	
1. Tetrahydrofuran/ 2,3-Dihydrofuran	10	60
2. 2-Methyltetrahydrofuran/ Tetrahydrofuran	10	60
3. Water/2,5-Dihydrofuran	60	200
4. Diethanolamine/Water	100	200
5. 1,2-Propanediol/ 1-Methoxy-2-propanol	75	120
6. Dipropylene glycol monomethyl ether/ 1,2-Propanediol	100	180
7. Acetophenone/Propene	25	60
8. Dimethyl disulfide/ Methanethiol	0	100

All the systems were studied by the PTx method described in the next section. System 3, water/2,5-dihydrofuran, was immiscible at 60°C; therefore, a solubility measurement was made at that temperature.

APPARATUS AND PROCEDURE

The PTx (total pressure-temperature-liquid composition) method is an efficient method of obtaining vapor-liquid equilibrium data. The required measurements are total pressure versus charge composition at constant temperature and known cell volume. With accurate pressure measurements and equations to model the vapor and liquid phases, PTx data can yield accurate phase composition information. An equation of state such as the Soave equation [1] is used to represent the nonidealities in the vapor phase and an activity coefficient equation such as the Wilson [2] or NRTL [3] equation is used to represent the nonidealities in the liquid phase.

Each system studied by the PTx method was studied along two isotherms across the entire composition range. Each isotherm was traversed in two or more runs. To initiate a run, the cell was charged with a known amount of one component. The cell contents were then degassed by removing vapor into a weighed evacuated cell or sample train. The measurement cell was allowed to equilibrate and the pressure was measured. Further degassing was performed until a repeatable vapor pressure was obtained. Increments of the second component were then charged to the cell. After each increment, the cell contents were degassed and allowed to equilibrate before the pressure was measured. The second and subsequent runs were similar to the first except that the second component was charged to the cell before adding increments of the first component. The ranges of compositions covered in the runs were designed to overlap to ensure consistency between the runs.

Systems 1,2 and the lower isotherm of system 4 were studied in the glass cell shown in Figure 1. The cell is made of thick-walled pyrex with a teflon cap and has an internal volume of 300 cm^3. The cap screws into the cell and forms a seal with an o-ring. One line through the cap is for adding components, the second line is for degassing, and the third is a thermowell into which a platinum resistance thermometer is inserted. A mercury manometer extends from the side of the cell and is connected to an exterior McLeod gauge for cell pressures below about 50 kPa, or to room pressure for higher pressures. The mercury levels in the cell manometer were read with a cathetometer. Pressures were measured to an accuracy of ±0.05 kPa. Temperatures in this and the other apparatuses were measured to an accuracy of ±0.05 K.

Systems 5 and 6 were studied using the glass still shown in Figure 2. The cell is connected to a large ballast tank and to pressure measurement equipment. Lines through the top of the cell are for charging and degassing. A thermowell extends through the top of the cell and into the liquid in the cell. The constant temperature bath in which the cell is placed was controlled at a temperature about 2°C warmer than the saturation temperature of the material in the cell. This caused the contents of the cell to reflux. The contents of the cell were vigorously stirred to ensure good contact and promote equilibrium between the vapor and liquid phases. The pressure in the ballast tank determined the temperature in the cell. The ballast pressure was adjusted until the desired cell temperature was attained. The cell pressure was measured with a manometer connected to a McLeod gauge or to room pressure. The pressure measurements in this apparatus were accurate to ±0.05 kPa.

The higher isotherm of system 3 and systems 7 and 8 were studied in the stainless steel cell shown in Figure 3. The cell has a volume of 300 cm^3 and is connected to external pressure gauges or manometers depending on the pressure range. There are lines entering the cell for charging, sampling and degassing. A thermowell also extends into the cell into which a platinum resistance thermometer is inserted. The cell and its connections are attached to a rigid support and immersed in a constant temperature bath. The cell was manually agitated to ensure the contents were at equilibrium at the desired temperature. Pressures were accurate to ±0.25% when measured with pressure gauges and ±0.05 kPa when measured with a manometer.

The lower isotherm of system 3 was measured in the apparatus shown in Figure 4. This apparatus is a glass flask with an attached mercury manometer much like the apparatus in Figure 1, but this apparatus has a much larger volume (600 cm^3) permitting large samples to be taken for analysis. The system was immiscible at this temperature, and samples were taken and analyzed by Karl-Fischer titration in order to determine the composition of the equilibrium liquid phases.

DATA REDUCTION PROCEDURE

Total pressure and charge composition data at constant temperature and cell volume may be reduced to equilibrium vapor and liquid compositions by use of an equation of state to model vapor phase nonidealities and an activity coefficient equation to model liquid phase nonidealities. These equations are used in conjunction with the basic equation of phase equilibrium to calculate vapor and liquid compositions from the measured total pressure data. The basic premise in this procedure is that the equation of state and the activity coefficient equation can adequately predict the behavior of the system.

The fundamental equation of phase equilibrium requires the fugacity of the ith component in a mixture to be equal in each phase.

$$f_i^\alpha = f_i^\beta \qquad (1)$$

where f_i is the fugacity of component i, and α, β represent separate phases. Equation (1) may be rewritten for vapor-liquid equilibrium as follows:

$$P y_i \phi_i = x_i \gamma_i P_i^\circ \phi_i^\circ \exp\left[\left(\frac{\bar{V}_i}{RT}\right)(P - P_i^\circ)\right] \qquad (2)$$

where P is the total pressure, y_i is the vapor mole fraction of component i, ϕ_i is the fugacity coefficient of component i in the vapor, x_i = liquid mole fraction of component i, γ_i is the activity coefficient of component i, P_i° is the vapor pressure of component i at system temperature, ϕ_i° is the fugacity coefficient of component i at system temperature and corresponding vapor pressure of component i, and the exponential term is the Poynting correction where $\bar{V}_i$ is the partial molar volume of component i. This volume is approximated from pure component density data from the literature adjusted to the temperature of the measurements using the Rackett equation [4].

The data reduction procedure consists of fitting the pressure data to Equation (2) across the entire composition range by adjusting the parameters of the activity coefficient model. The Wilson and NRTL equations were both used in this work. These equations are listed below:

Wilson equation

$$\ln \gamma_1 = -\ln(x_1 + \Lambda_{12} x_2) + x_2 \left[\frac{\Lambda_{12}}{x_1 + \Lambda_{12} x_2} - \frac{\Lambda_{21}}{\Lambda_{21} x_1 + x_2}\right] \qquad (3)$$

$$\ln \gamma_2 = -\ln(x_2 + \Lambda_{21} x_1) - x_1 \left[\frac{\Lambda_{12}}{x_1 + \Lambda_{12} x_2} - \frac{\Lambda_{21}}{\Lambda_{21} x_1 + x_2}\right] \qquad (4)$$

NRTL equation

$$\ln \gamma_1 = x_2^2 \left[\tau_{21}\left(\frac{G_{21}}{x_1 + x_2 G_{21}}\right)^2 + \left(\frac{\tau_{12} G_{12}}{(x_2 + x_1 G_{12})^2}\right)\right] \qquad (5)$$

$$\ln \gamma_2 = x_1^2 \left[\tau_{12}\left(\frac{G_{12}}{x_2 + x_1 G_{12}}\right)^2 + \left(\frac{\tau_{21} G_{21}}{(x_1 + x_2 G_{21})^2}\right)\right] \qquad (6)$$

$$G_{12} = \exp(-\alpha_{12} \tau_{12}), \quad G_{21} = \exp(-\alpha_{21} \tau_{21}) \qquad (7)$$

An initial estimate of the solution of Equation (2) for each component is obtained by assuming that the liquid composition is the same as the charge composition, the fugacity coefficients and Poynting correction are equal to unity, and by using the ideal parameter values from the activity coefficient model. Then Equation (2) is solved for the product $P y_i$ for each component. The calculated pressure is then the sum of these products:

$$P_{calc} = \sum_i (P y_i) \qquad (8)$$

The vapor mole fraction for each component is then determined:

$$y_i = (P y_i)/P_{calc} \qquad (9)$$

With values for the vapor-phase composition, the fugacity coefficients can be calculated from the equation of state.

The next step is to correct the liquid composition for the amounts of each component in the vapor, and to return to the step in which the activity coefficients are calculated and continue iterating until the calculated pressure converges. As part of each iteration step, the amount of material taken out of the cell as degas is subtracted from the total charge at the calculated vapor composition.

This procedure is performed for each of the measurement points across the composition range. The calculated pressures are compared to the measured pressures and the activity coefficient parameters are adjusted to improve the fit of the total pressure data. The entire procedure is repeated until the calculated pressure curve agrees well with the measured pressure points.

The critical constants and other properties used in the data reduction procedure are given in Table 1.

RESULTS AND DISCUSSION

The results of the data reduction procedure are the calculated equilibrium phase compositions, a comparison of the measured and calculated pressures across the entire composition range, the calculated activity coefficients and the parameters of the activity coefficient model used to fit the measured data. These results are given for each of the PTx Isotherms described below. The RMS deviation between the measured and correlated pressure is given for each PTx run.

Tetrahydrofuran/2,3-Dihydrofuran

Tables 2 and 3 give the results of the PTx measurements on the tetrahydrofuran/2,3-dihydrofuran system at 10 and 60°C. Nearly ideal behavior is exhibited with only slight negative deviations. The NRTL parameters used to correlate the data are given at the bottom of each table. Figure 5 shows the measured data and the total-pressure correlation at 10°C, and Figure 6 shows the derived y-x plot. Figure 7 is the total-pressure plot for the measurements at 60°C, and Figure 8 is the derived y-x plot for these data.

2-Methyltetrahydrofuran/Tetrahydrofuran

Tables 4 and 5 present the results for the 2-methyltetrahydrofuran/tetrahydrofuran system. Because the two compounds are so nearly alike the behavior of this system is ideal. At the lower temperature of 10°C both infinite dilution activity coefficients are 1.012, indicating ideality. At 60°C complete ideality was observed. Figures 9 and 10 illustrate these results for the 10°C measurements, and Figures 11 and 12 illustrate these results for the 60°C measurements.

Water/2,5-Dihydrofuran

Originally, it was intended that the water/2,3-dihydrofuran binary be studied, but there was significant reaction between water and 2,3-dihydrofuran so the DIPPR 805 committee substituted the water/2,5-dihydrofuran system. This system is immiscible at room temperature and at temperatures up to near 200°C. One set of liquid-liquid equilibrium measurements were made at 60°C and a PTx run was made at 200°C.

Table 6 shows the results of the LLE measurements at 60°C. Table 7 contains the results of the PTx measurements at 200°C. Figures 13 and 14 show the results of the PTx measurements and show the unusual behavior exhibited by this system. At 200°C a minimum boiling azeotrope occurs at approximately 53 mole % 2,5-dihydrofuran.

Diethanolamine/Water

The diethanolamine/water system was studied at 100 and 200°C. The PTx results are given in Tables 8 and 9. Slight negative deviation from Raoult's law is exhibited as is seen in Figures 15 and 17 which are the total-pressure plots for these measurements. Because of the large difference in vapor pressure, the relative volatility of water over diethanolamine is very large, as shown in the tables and in the y-x plots which are Figures 16 and 18.

1,2-Propanediol/1-Methoxy-2-Propanol

Tables 10 and 11 contain the PTx results for the 1,2-propanediol/1-methoxy-2-propanol system. Slight positive deviation from ideality is exhibited at both temperatures of 75 and 120°C. The NRTL equation was used to model these data, and the parameters are given at the bottom of the tables. Figures 19 and 21 show the experimental results and the correlated pressure curves for each isotherm and Figures 20 and 22 show the y-x plots.

1,2-Propanediol/Dipropylene Glycol Monomethyl Ether

The results of PTx measurements on the 1,2-propanediol/dipropylene glycol monomethyl ether binary system at 100 and 180°C are given in Tables 12 and 13. This system showed pronounced positive deviation from ideality, and because the vapor pressures of the two pure compounds are nearly the same, a minimum boiling azeotrope occurs at both temperatures studied. At 100°C dipropylene glycol monomethyl ether has a higher vapor pressure, but at 180°C 1,2-propanediol has a higher vapor pressure. The Wilson equation was to reduce the PTx data obtained for this system. The parameters are given in the tables.

Figures 23 and 24 show the total pressure plot and y-x plot, respectively, for the measurements at 100°C, and Figures 25 and 26 show these plots for the 180°C results.

Acetophenone/Propene

The acetophenone/propene system was studied at 25 and 60°C. The results are given in Tables 14 and 15 and Figures 27 through 30. The data were modeled with the NRTL equation. Significant positive deviation from ideality was observed at both temperatures, and because of the large difference in pure component vapor pressures very large relative volatilities are witnessed.

Dimethyl Disulfide/Methanethiol

The dimethyl disulfide/methanethiol (methyl mercaptan) system was studied at 0 and 100°C. This system exhibited nearly ideal behavior as shown in Tables 16 and 17. The activity coefficients show a minimum for dimethyl disulfide and a maximum for methanethiol. The Wilson equation parameters used to obtain these results are given at the bottom of each table. Figures 31 through 34 show the results as total pressure plots and y-x diagrams.

Vapor Pressures

As part of each PTx run the pure component vapor pressure of the first component charged to the measurement cell is measured. Table 18 shows the measured vapor pressures for the pure components compared to various literature sources where available.

Table 19 shows the source and purity of the chemicals used in the measurements.

ACKNOWLEDGEMENT

We express our appreciation to the American Institute of Chemical Engineers and to the DIPPR 805 committee for the opportunity to participate in DIPPR 805 projects.

LITERATURE CITED

1. Soave, G., *Chem. Eng. Sci.*, **27**, 1197 (1972).

2. Wilson, G.M. and G. Scatchard, *J. Am. Chem. Soc.*, **86**, 125 (1964).

3. Renon, H. and J.M. Prausnitz, *AIChE J.*, **14**, 135 (1968).

4. Rackett, H.G., *J. Chem. Eng. Data*, **15**, 514 (1970).

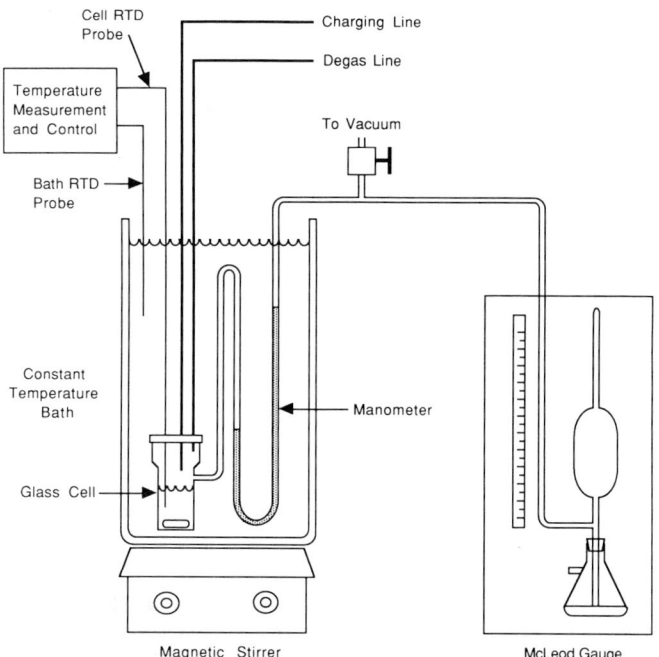

Figure 1. Glass PT$_x$ Apparatus

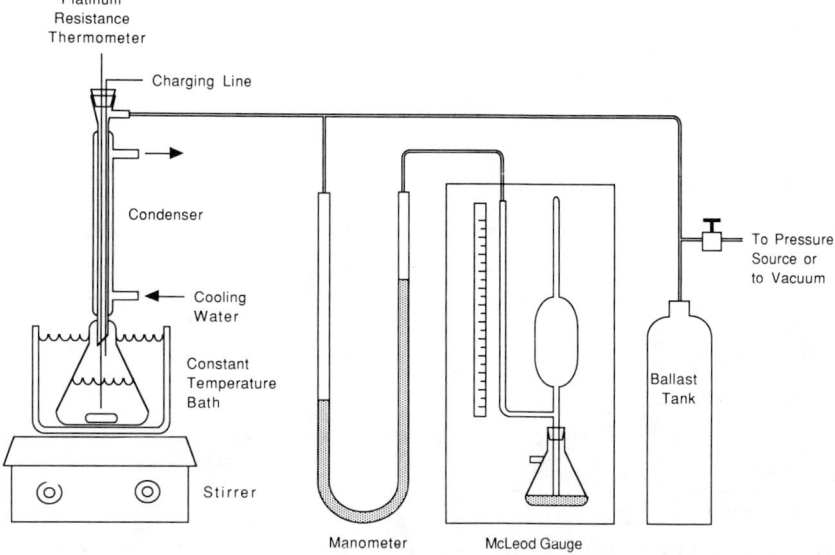

Figure 2. Glass Still PT$_x$ Apparatus

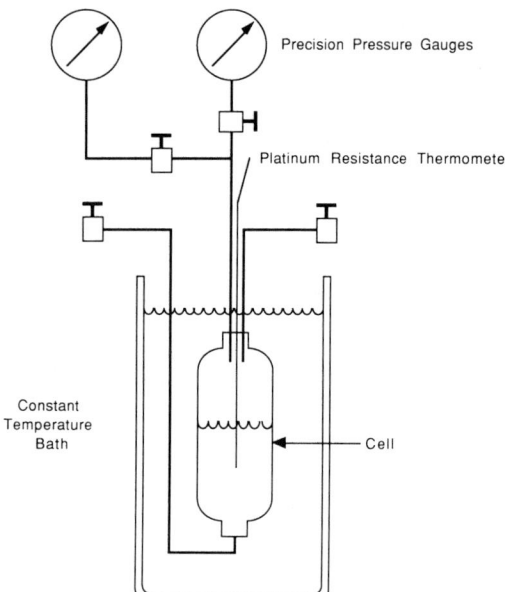

Figure 3. Stainless Steel PT$_X$ Apparatus

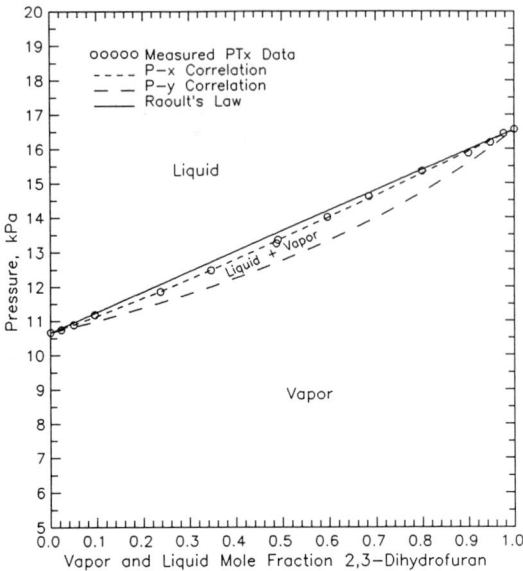

Figure 5. Measured Pressure versus Calculated Vapor and Liquid Compositions for the Tetrahydrofuran/2,3-Dihydrofuran System at 10°C

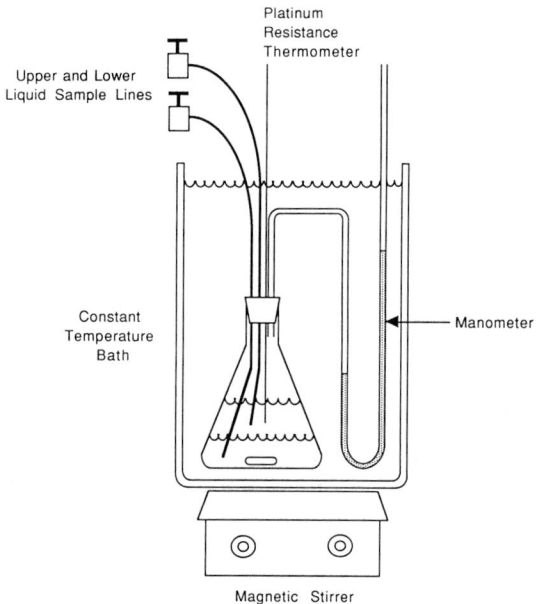

Figure 4. Large-Volume, Liquid-Liquid Equilibrium Apparatus

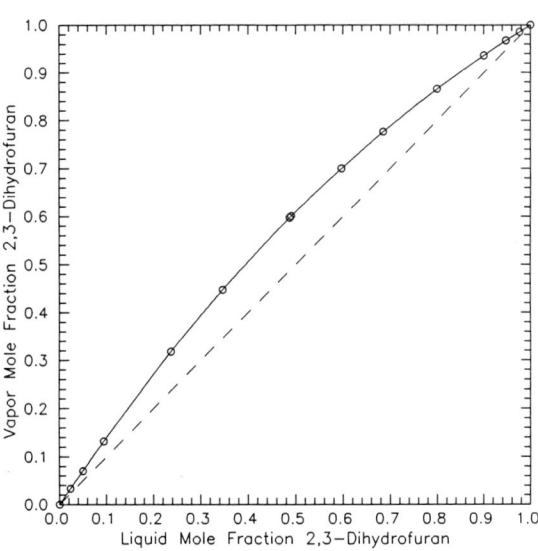

Figure 6. Y-X Diagram for the Tetrahydrofuran/2,3-Dihydrofuran System at 10°C

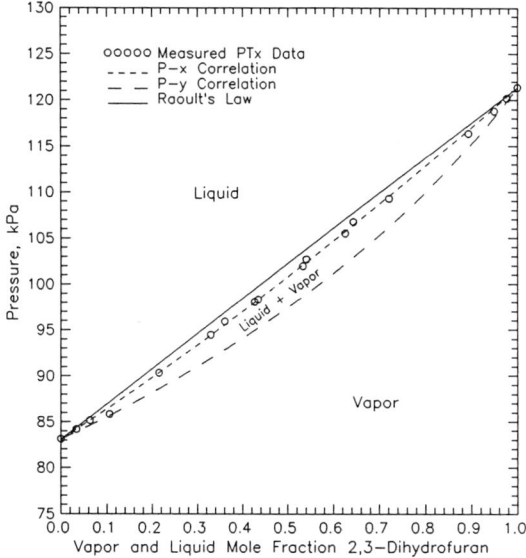

Figure 7. Measured Pressure versus Calculated Vapor and Liquid Compositions for the Tetrahydrofuran/2,3-Dihydrofuran System at 60°C

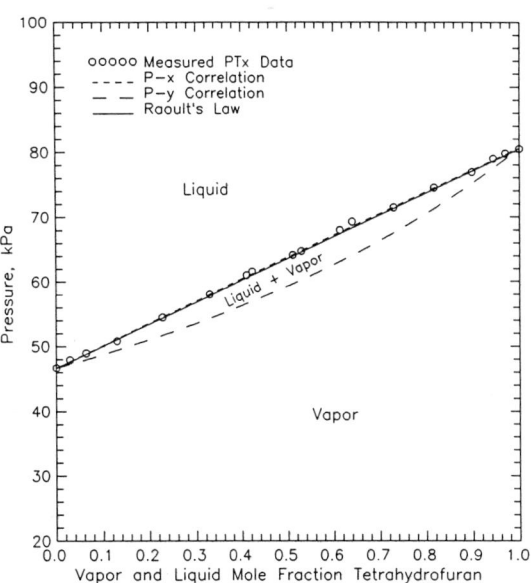

Figure 9. Measured Pressure versus Calculated Vapor and Liquid Compositions for the 2-Methyltetrahydrofuran/Tetrahydrofuran System at 10°C

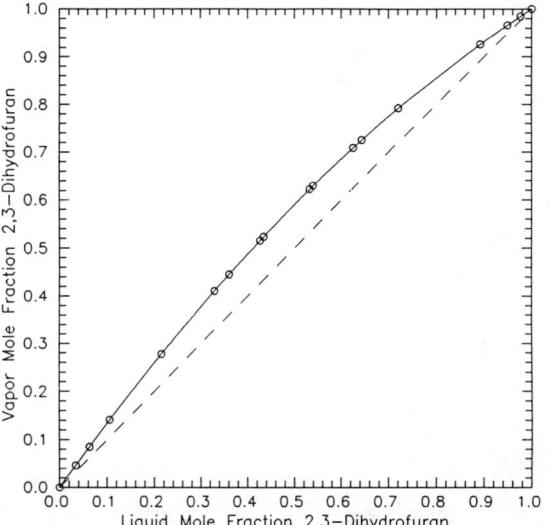

Figure 8. Y-X Diagram for the Tetrahydrofuran/2,3-Dihydrofuran System at 60°C

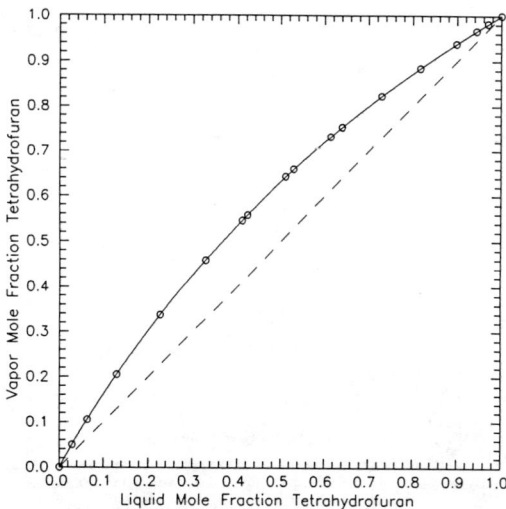

Figure 10. Y-X Diagram for the 2-Methyltetrahydrofuran/Tetrahydrofuran System at 10°C

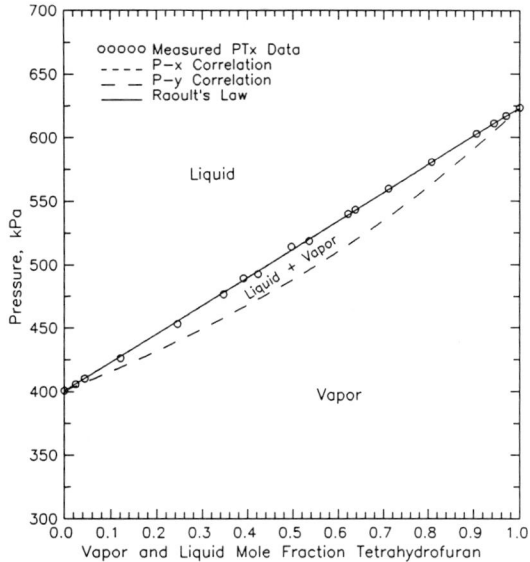

Figure 11. Measured Pressure versus Calculated Vapor and Liquid Compositions for the 2-Methyltetrahydrofuran/Tetrahydrofuran System at 60°C

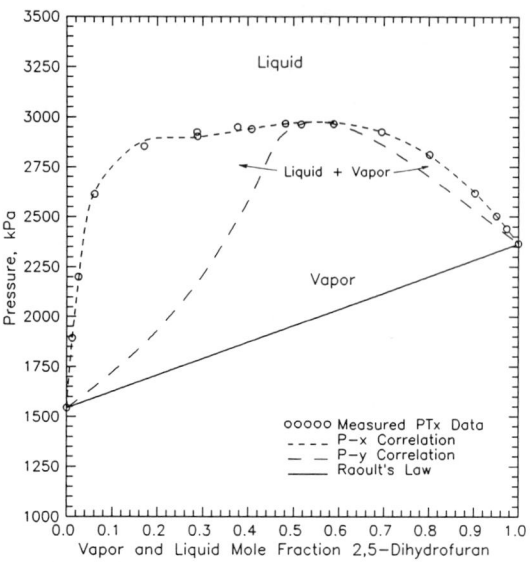

Figure 13. Measured Pressure versus Calculated Vapor and Liquid Compositions for the Water/2,5-Dihydrofuran System at 200°C

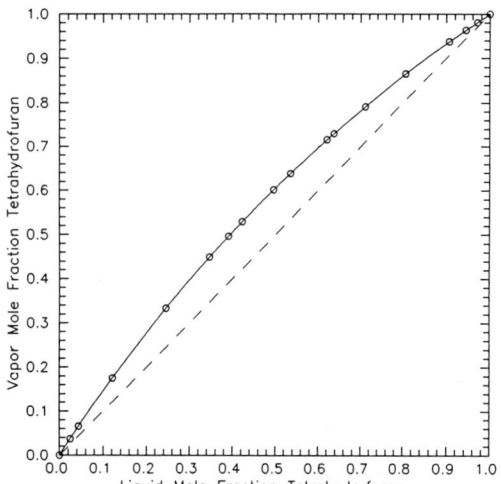

Figure 12. Y-X Diagram for the 2-Methyltetrahydrofuran/Tetrahydrofuran System at 60°C

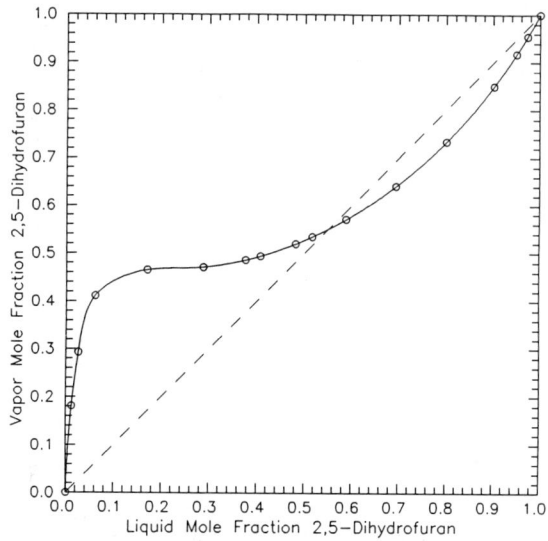

Figure 14. Y-X Diagram for the Water/2,5-Dihydrofuran System at 200°C

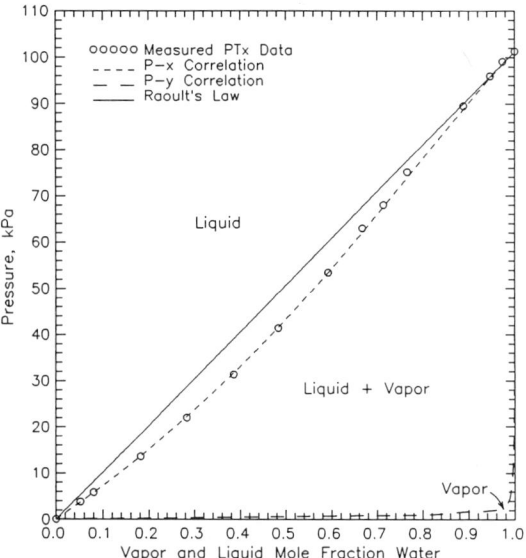

Figure 15. Measured Pressure versus Calculated Vapor and Liquid Compositions for the Diethanolamine/Water System at 100°C

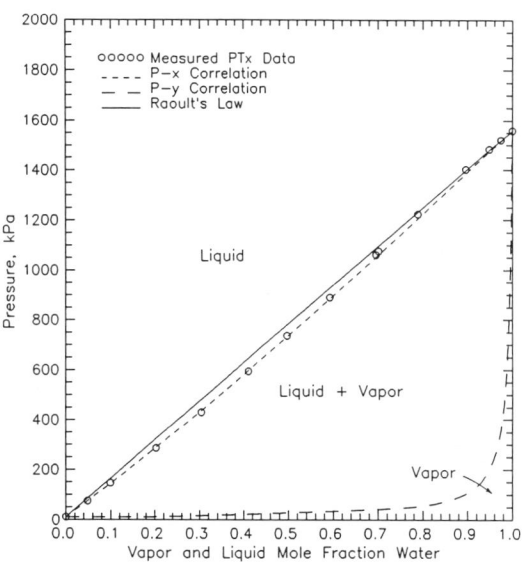

Figure 17. Measured Pressure versus Calculated Vapor and Liquid Compositions for the Diethanolamine/Water System at 200°C

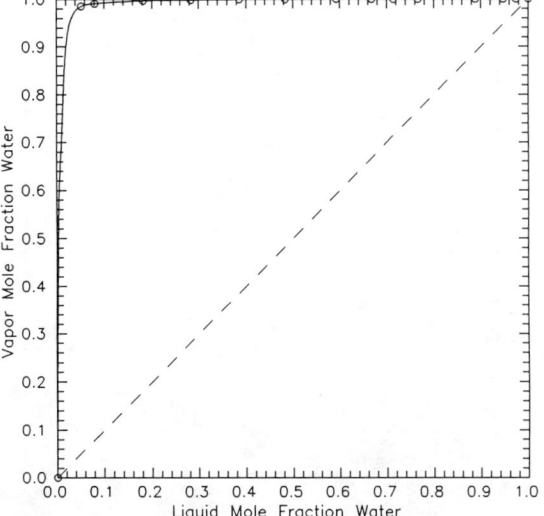

Figure 16. Y-X Diagram for the Diethanolamine/Water System at 100°C

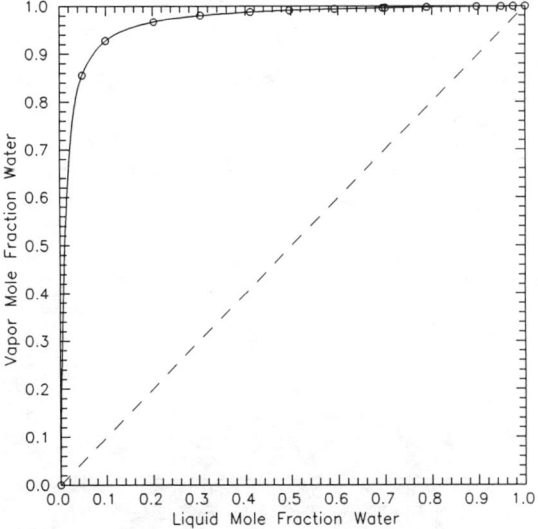

Figure 18. Y-X Diagram for the Diethanolamine/Water System at 200°C

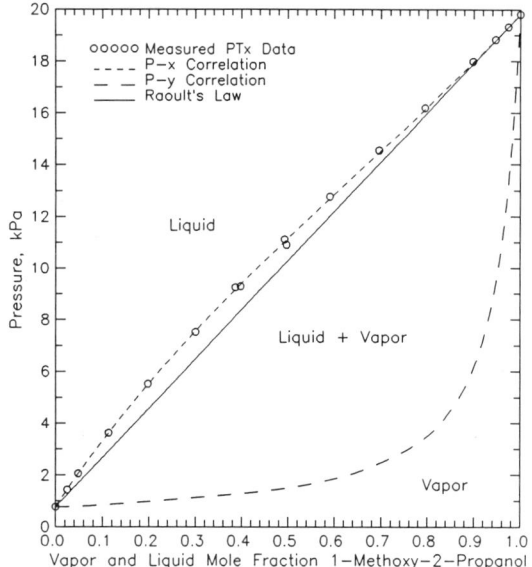

Figure 19. Measured Pressure versus Calculated Vapor and Liquid Compositions for the 1,2-Propanediol/1-Methoxy-2-Propanol System at 75°C

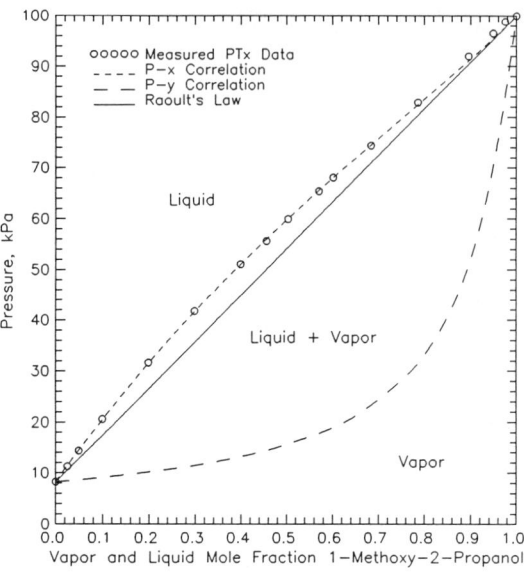

Figure 21. Measured Pressure versus Calculated Vapor and Liquid Compositions for the 1,2-Propanediol/1-Methoxy-2-Propanol System at 120°C

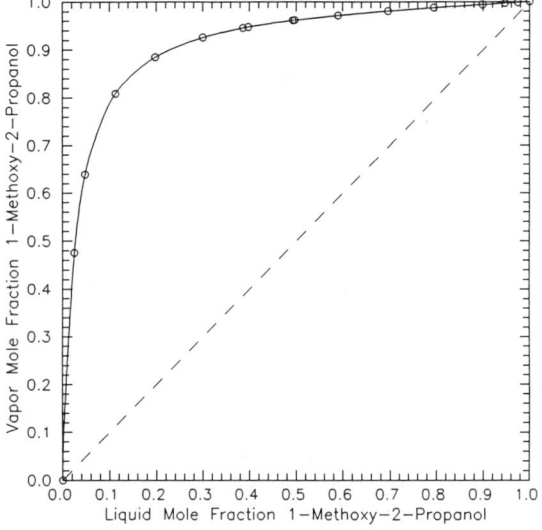

Figure 20. Y-X Diagram for the 1,2-Propanediol/1-Methoxy-2-Propanol System at 75°C

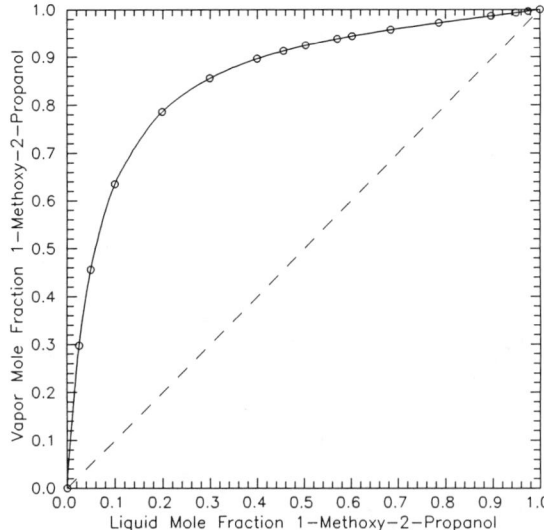

Figure 22. Y-X Diagram for the 1,2-Propanediol/1-Methoxy-2-Propanol System at 120°C

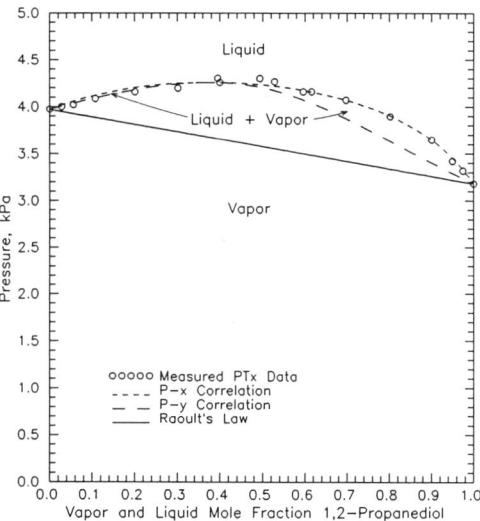

Figure 23. Measured Pressure versus Calculated Vapor and Liquid Compositions for the 1,2-Propanediol/Dipropylene Glycol Monomethyl Ether System at 100°C

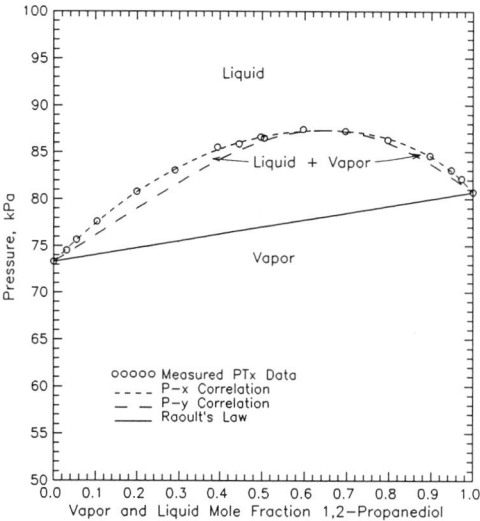

Figure 25. Measured Pressure versus Calculated Vapor and Liquid Compositions for the 1,2-Propanediol/Dipropylene Glycol Monomethyl Ether System at 180°C

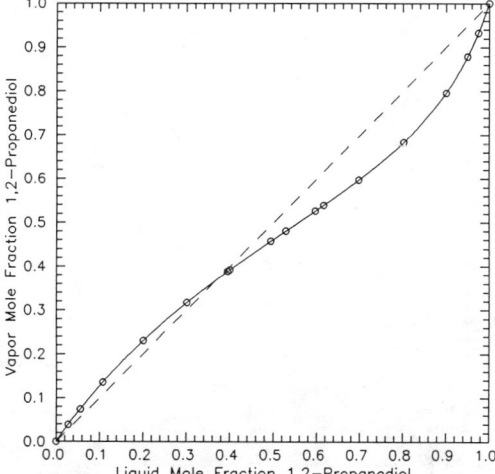

Figure 24. Y-X Diagram for the 1,2-Propanediol/Dipropylene Glycol Monomethyl Ether System at 100°C

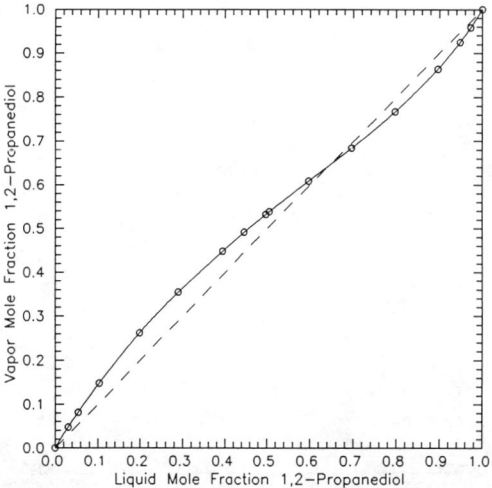

Figure 26. Y-X Diagram for the 1,2-Propanediol/Dipropylene Glycol Monomethyl Ether System at 180°C

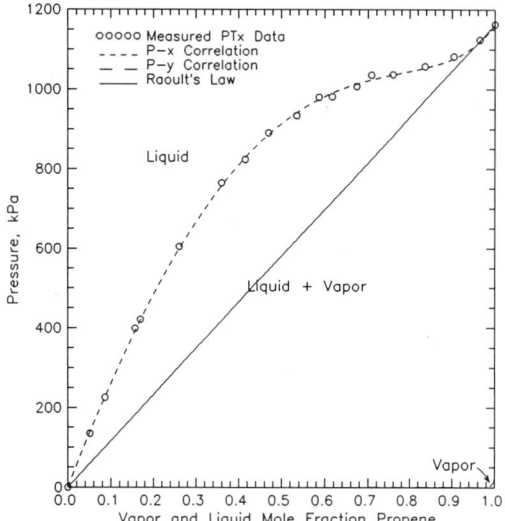

Figure 27. Measured Pressure versus Calculated Vapor and Liquid Compositions for the Acetophenone/Propene System at 25°C

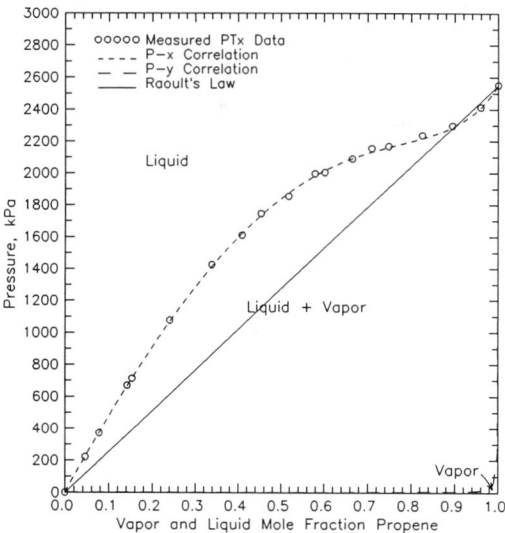

Figure 29. Measured Pressure versus Calculated Vapor and Liquid Compositions for the Acetophenone/Propene System at 60°C

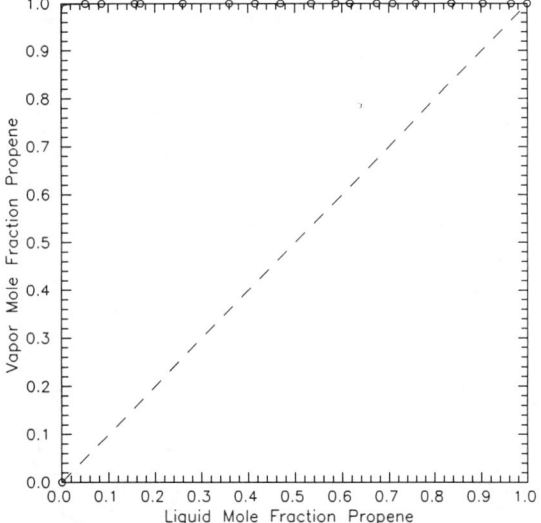

Figure 28. Y-X Diagram for the Acetophonone/Propene System at 25°C

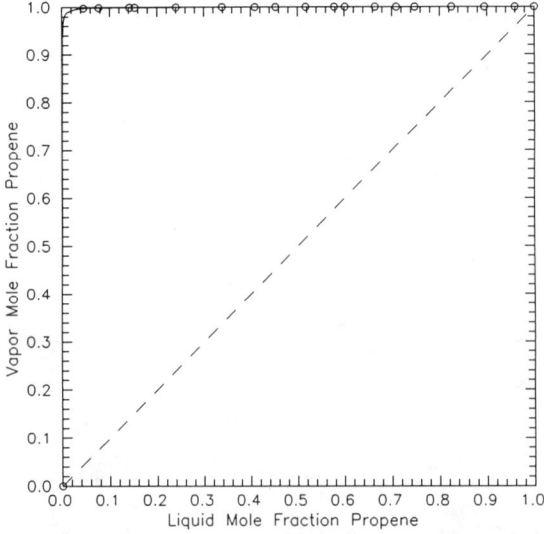

Figure 30. Y-X Diagram for the Acetophenone/Propene System at 60°C

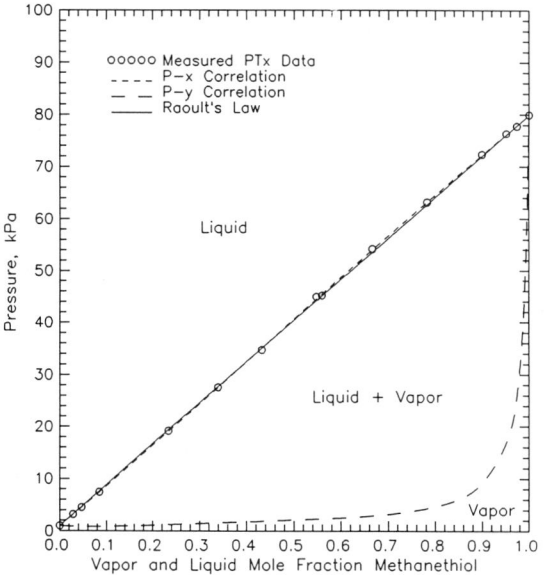

Figure 31. Measured Pressure versus Calculated Vapor and Liquid Compositions for the Dimethyl Disulfide/Methanethiol System at 0°C

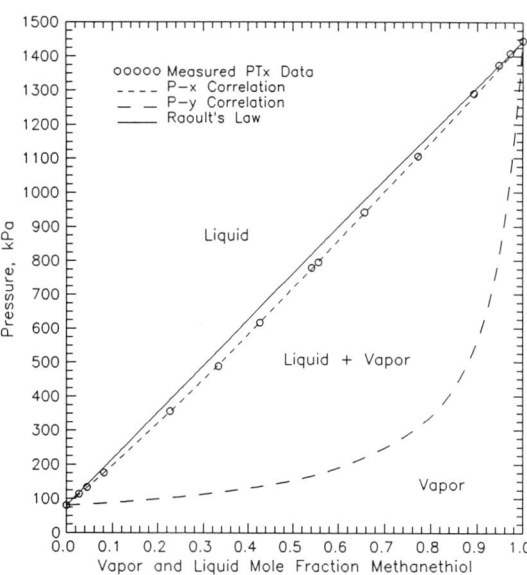

Figure 33. Measured Pressure versus Calculated Vapor and Liquid Compositions for the Dimethyl Disulfide/Methanethiol System at 100°C

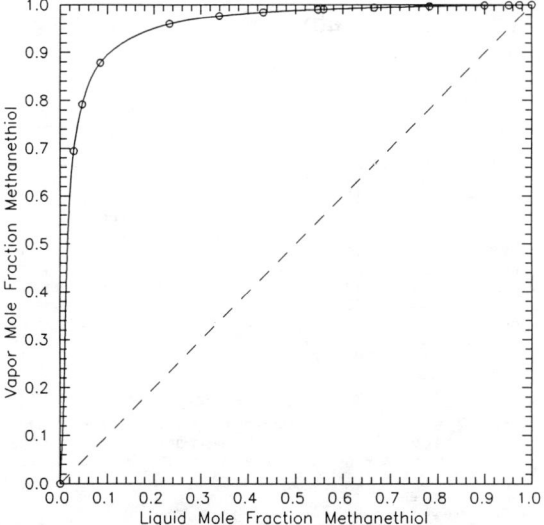

Figure 32. Y-X Diagram for the Dimethyl Disulfide/Methanethiol System at 0°C

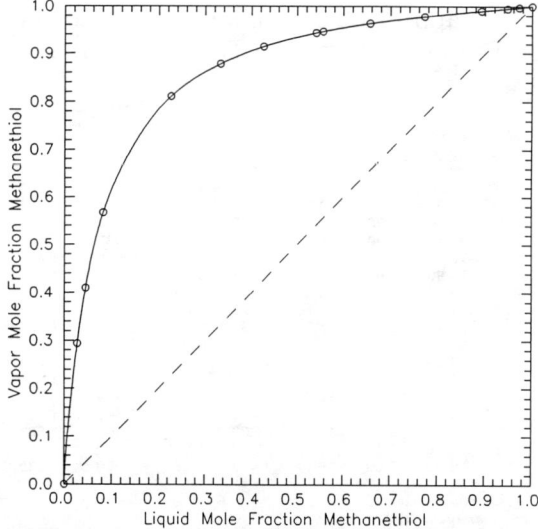

Figure 34. Y-X Diagram for the Dimethyl Disulfide/Methanethiol System at 100°C

Table 1. Constants and Properties Used in Data Reduction

	Mol Wt	Tc, K	Pc, kPa	Zc	Acentric Factor	Liquid Density g/cc	Liquid Density Temp, K
Tetrahydrofuran	72.107	540.1	5190	0.259	0.217	0.887	293
2,3-Dihydrofuran	70.091	515.15	5500	0.277	0.213	0.927	293
2-Methyltetrahydrofuran	86.134	537.0	3711	0.225	0.264	0.860	293
2,5-Dihydrofuran	70.091	541.39	5391	0.250	0.2184	0.927	298
Water	18.015	647.3	22120	0.229	0.344	1.000	298
Diethanolamine	105.14	708.66	5066	0.286	1.301	1.097	298
1,2-Propanediol	76.096	625.60	6070	0.280	1.116	1.036	298
1-Methoxy-2-propanol	148.55	554.43	4334	0.273	0.705	0.922	298
Dipropylene glycol monomethyl ether	80.148	620.14	3032	0.277	0.876	0.950	298
Acetophenone	120.15	714.0	4060	0.257	0.420	1.032	288
Propene	42.081	364.9	4600	0.274	0.144	0.612	223
Dimethyl disulfide	94.189	606.3	5280	0.275	0.258	1.046	298
Methanethiol	48.107	470.0	7235	0.268	0.155	0.866	293

Table 2. Vapor-Liquid Equilibrium Measurements on the Tetrahydrofuran/2,3-Dihydrofuran System at 10°C

Run Num	Mole Percent THF Charge	Mole Percent THF Liquid	Mole Percent THF Vapor	Pressure, kPa Meas	Pressure, kPa Calc	Activity Coefficient THF	Activity Coefficient 23-DHF	Relative Volatility 23-DHF/THF
1	100.00	100.00	100.00	10.67	10.67	1.000	0.916	1.420
1	97.68	97.68	96.72	10.75	10.78	1.000	0.922	1.428
1	95.00	95.00	92.96	10.89	10.90	1.000	0.928	1.438
1	90.52	90.53	86.80	11.18	11.12	0.999	0.937	1.454
1	76.26	76.28	68.22	11.85	11.86	0.994	0.961	1.498
2	70.90	70.90	61.69	12.44	12.16	0.991	0.968	1.513
1	65.33	65.34	55.25	12.47	12.48	0.988	0.974	1.527
2	61.05	61.05	50.48	12.90	12.73	0.986	0.978	1.538
1	51.24	51.25	40.27	13.25	13.32	0.980	0.986	1.559
2	50.91	50.92	39.94	13.35	13.34	0.980	0.986	1.560
2	40.34	40.35	29.97	14.01	13.99	0.973	0.992	1.581
2	31.48	31.50	22.37	14.61	14.55	0.967	0.996	1.596
2	20.02	20.03	13.44	15.34	15.28	0.959	0.998	1.613
2	9.98	9.99	6.39	15.87	15.93	0.953	1.000	1.625
2	5.22	5.23	3.27	16.19	16.23	0.950	1.000	1.631
2	2.35	2.35	1.45	16.45	16.42	0.948	1.000	1.634
2	0.00	0.00	0.00	16.57	16.57	0.947	1.000	1.636

NRTL Equation Parameters used to Fit Measurements
$\tau_{12} = -0.2436$
$\tau_{21} = 0.1363$
$\alpha = -1.00$

RMS Deviation = 0.26%

Table 3. Vapor-Liquid Equilibrium Measurements on the Tetrahydrofuran/2,3-Dihydrofuran System at 60°C

Run Num	Mole Percent THF Charge	Mole Percent THF Liquid	Mole Percent THF Vapor	Pressure, kPa Meas	Pressure, kPa Calc	Activity Coefficient THF	Activity Coefficient 23-DHF	Relative Volatility 23-DHF/THF
1	100.00	100.00	100.00	83.13	83.13	1.000	0.948	1.370
1	96.62	96.63	95.43	84.19	84.20	1.000	0.952	1.375
1	93.69	93.72	91.54	85.16	85.14	1.000	0.954	1.379
1	89.39	89.43	85.93	85.83	86.56	0.999	0.958	1.386
1	78.40	78.47	72.23	90.32	90.29	0.998	0.968	1.402
1	66.98	67.06	58.94	94.45	94.37	0.994	0.976	1.419
2	63.88	63.92	55.45	95.91	95.52	0.993	0.978	1.423
1	57.26	57.34	48.40	98.05	97.98	0.990	0.983	1.433
2	56.51	56.57	47.59	98.29	98.28	0.990	0.983	1.434
2	46.65	46.73	37.70	101.94	102.07	0.985	0.988	1.449
1	45.88	45.94	36.95	102.67	102.37	0.985	0.989	1.450
2	37.37	37.46	29.04	105.50	105.74	0.979	0.993	1.463
1	35.61	35.65	27.43	106.74	106.47	0.978	0.993	1.466
2	27.85	27.94	20.78	109.26	109.60	0.973	0.996	1.478
2	10.72	10.78	7.43	116.37	116.77	0.958	0.999	1.505
2	5.08	5.12	3.44	118.80	119.18	0.953	1.000	1.514
2	2.38	2.39	1.59	120.19	120.35	0.950	1.000	1.518
2	0.00	0.00	0.00	121.38	121.38	0.948	1.000	1.522

NRTL Equation Parameters used to Fit Measurements
$\tau_{12} = -0.0258$
$\tau_{21} = -0.0282$
$\alpha = -1.00$

RMS Deviation = 0.36%

Table 4. Vapor-Liquid Equilibrium Measurements on the 2-Methyltetrahydrofuran/Tetrahydrofuran System at 10°C

Run Num	Mole Percent mTHF Charge	Mole Percent mTHF Liquid	Mole Percent mTHF Vapor	Pressure, kPa Meas	Pressure, kPa Calc	Activity Coefficient mTHF	Activity Coefficient mTHF	Relative Volatility THF/mTHF
1	100.00	100.00	100.00	46.68	46.68	1.000	1.012	1.742
1	97.08	97.09	95.03	47.92	47.69	1.000	1.011	1.741
1	93.65	93.66	89.46	48.97	48.88	1.000	1.010	1.740
1	87.03	87.04	79.45	50.81	51.17	1.000	1.009	1.737
1	77.27	77.28	66.25	54.45	54.52	1.001	1.007	1.733
1	67.11	67.12	54.15	57.99	57.99	1.001	1.005	1.728
1	58.90	58.91	45.39	60.98	60.78	1.002	1.004	1.725
2	57.69	57.69	44.16	61.53	61.19	1.002	1.004	1.724
1	48.85	48.85	35.70	64.12	64.18	1.003	1.003	1.721
2	46.99	47.00	34.02	64.72	64.80	1.003	1.003	1.720
1	38.60	38.61	26.82	67.96	67.63	1.004	1.002	1.716
2	35.98	35.99	24.69	69.26	68.51	1.005	1.002	1.715
2	27.01	27.02	17.78	71.45	71.52	1.006	1.001	1.712
2	18.30	18.31	11.60	74.50	74.43	1.008	1.000	1.708
2	10.13	10.14	6.21	76.89	77.16	1.010	1.000	1.705
2	5.60	5.61	3.37	78.94	78.67	1.011	1.000	1.703
2	3.00	3.00	1.78	79.74	79.54	1.011	1.000	1.702
2	0.00	0.00	0.00	80.45	80.54	1.012	1.000	1.700

NRTL Equation Parameters used to Fit Measurements
$\tau_{12} = 0.00595$
$\tau_{21} = 0.00595$
$\alpha = -1.00$

RMS Deviation = 0.36%

Table 5. Vapor-Liquid Equilibrium Measurements on the 2-Methyltetrahydrofuran/Tetrahydrofuran System at 60°C

Run Num	Mole Percent mTHF Charge	Mole Percent mTHF Liquid	Mole Percent mTHF Vapor	Pressure, kPa Meas	Pressure, kPa Calc	Activity Coefficient mTHF	Activity Coefficient mTHF	Relative Volatility THF/mTHF
1	100.00	100.00	100.00	400.9	400.9	1.000	1.000	1.540
1	97.56	97.57	96.30	406.0	406.3	1.000	1.000	1.540
1	95.62	95.64	93.44	410.4	410.6	1.000	1.000	1.540
1	87.81	87.86	82.45	426.3	427.8	1.000	1.000	1.539
1	75.42	75.49	66.68	453.1	455.3	1.000	1.000	1.539
1	65.26	65.32	55.04	476.3	477.8	1.000	1.000	1.539
2	60.90	60.93	50.33	489.2	487.6	1.000	1.000	1.539
1	57.66	57.73	47.02	492.5	494.7	1.000	1.000	1.538
2	50.33	50.38	39.76	514.0	511.1	1.000	1.000	1.538
1	46.43	46.48	36.09	518.5	519.7	1.000	1.000	1.538
2	37.83	37.89	28.40	539.7	538.9	1.000	1.000	1.538
1	36.24	36.27	27.01	543.1	542.5	1.000	1.000	1.538
2	28.84	28.91	20.92	559.7	558.9	1.000	1.000	1.538
2	19.27	19.32	13.48	580.3	580.3	1.000	1.000	1.537
2	9.30	9.34	6.28	602.7	602.5	1.000	1.000	1.537
2	5.52	5.54	3.67	611.0	611.0	1.000	1.000	1.537
2	2.92	2.93	1.92	616.9	616.8	1.000	1.000	1.537
2	0.00	0.00	0.00	623.4	623.4	1.000	1.000	1.537

NRTL Equation Parameters used to Fit Measurements
$\tau_{12} = 0.000$
$\tau_{21} = 0.000$
$a = -1.00$

RMS Deviation = 0.22%

Table 6. Liquid-Liquid Equilibrium Measurements for the System Water/2,5-Dihydrofuran at 60°C

Measured Pressure: 87.98 kPa

	Mole % H2O	Mole % 25DHF	Weight % H2O	Weight % 25DHF
Upper Liquid	36.68±0.03	63.32±0.03	12.96±0.02	87.04±0.02
Lower Liquid	92.93±0.23	7.07±0.23	77.17±0.62	22.83±0.62

Table 7. Vapor-Liquid Equilibrium Measurements for the Water/2,5-Dihydrofuran System at 200°C

Run Num	Mole Percent Water Charge	Mole Percent Water Liquid	Mole Percent Water Vapor	Pressure, kPa Meas	Pressure, kPa Calc	Activity Coefficient Water	Activity Coefficient 25DHF	Relative Volatility 25DHF/Water
1	100.00	100.00	100.00	1544	1544	1.000	15.747	20.81
1	97.64	98.80	81.89	1892	1885	1.001	13.258	18.21
1	95.26	97.42	70.71	2200	2184	1.005	11.036	15.63
1	90.52	93.90	58.88	2613	2623	1.023	7.362	10.75
1	80.31	82.93	53.51	2852	2880	1.139	3.135	4.223
2	70.94	71.25	52.94	2923	2898	1.319	1.889	2.203
1	69.95	71.16	52.93	2901	2898	1.320	1.884	2.194
2	62.01	62.31	51.40	2948	2929	1.480	1.495	1.563
1	58.87	59.23	50.59	2938	2941	1.539	1.408	1.419
2	51.66	51.82	48.04	2965	2965	1.685	1.256	1.164
1	48.31	48.33	46.54	2961	2972	1.756	1.205	1.075
2	41.26	41.16	42.87	2964	2971	1.906	1.127	0.932
2	30.92	30.54	35.87	2924	2923	2.134	1.058	0.786
2	20.41	19.85	26.57	2810	2805	2.370	1.021	0.685
2	10.27	9.79	15.06	2616	2619	2.595	1.004	0.612
2	5.27	4.97	8.24	2502	2502	2.704	1.001	0.583
2	2.83	2.66	4.58	2439	2440	2.756	1.000	0.570
2	0.00	0.00	0.00	2364	2364	2.817	1.000	0.555

NRTL Equation Parameters used to Fit Measurements
$\tau_{12} = 0.0336$
$\tau_{21} = 1.0009$
$a = -1.00$

RMS Deviation = 0.42%

Table 8. Vapor-Liquid Equilibrium Measurements for the Diethanolamine/Water System at 100°C

Run Num	Mole Percent DEA Charge	Mole Percent DEA Liquid	Mole Percent DEA Vapor	Pressure, kPa Meas	Pressure, kPa Calc	Activity Coefficient DEA	Activity Coefficient Water	Relative Volatility Water/DEA
1	100.00	100.00	100.00	0.059	0.059	1.000	0.691	1182
1	94.78	94.80	1.49	3.79	3.74	0.999	0.707	1208
1	91.92	91.96	0.93	5.85	5.83	0.999	0.715	1223
1	81.80	81.86	0.35	13.56	13.67	0.992	0.747	1282
1	71.59	71.67	0.19	22.04	22.30	0.979	0.781	1355
1	61.37	61.45	0.11	31.36	31.74	0.957	0.817	1445
1	51.58	51.65	0.07	41.36	41.57	0.926	0.853	1553
1	40.73	40.78	0.04	53.42	53.35	0.877	0.893	1710
2	33.20	33.24	0.03	62.95	62.05	0.832	0.920	1850
1	28.57	28.61	0.02	67.99	67.56	0.800	0.937	1955
2	23.33	23.38	0.01	75.13	73.89	0.759	0.954	2094
2	11.12	11.15	0.00	89.52	88.77	0.643	0.987	2547
2	5.31	5.33	0.00	95.98	95.55	0.577	0.997	2858
2	2.60	2.61	0.00	99.12	98.56	0.544	0.999	3034
2	0.00	0.00	0.00	101.3	101.3	0.511	1.000	3228

NRTL Equation Parameters used to Fit Measurements
$\tau_{12} = -0.0320$
$\tau_{21} = -0.6339$
$a = -1.00$

RMS Deviation = 0.89%

Table 9. Vapor-Liquid Equilibrium Measurements for the Diethanolamine/Water System at 200°C

Run Num	Mole Percent DEA Charge	Mole Percent DEA Liquid	Mole Percent DEA Vapor	Pressure, kPa Meas	Pressure, kPa Calc	Activity Coefficient DEA	Activity Coefficient Water	Relative Volatility Water/DEA
1	100.00	100.00	100.00	11.29	11.29	1.000	0.910	116.3
1	94.91	95.15	14.45	75.15	75.47	1.000	0.918	116.1
1	89.65	90.09	7.28	146.31	144.04	0.999	0.927	115.8
1	79.02	79.72	3.30	286.13	289.34	0.996	0.942	115.1
1	68.81	69.61	1.97	429.54	436.74	0.991	0.955	114.2
1	58.16	58.93	1.25	594.33	598.33	0.984	0.968	113.2
1	49.67	50.33	0.89	736.36	732.14	0.977	0.976	112.3
1	40.30	40.78	0.62	889.42	884.10	0.968	0.984	111.3
2	30.35	30.49	0.40	1060.4	1051.3	0.956	0.991	110.0
1	29.65	29.92	0.39	1074.9	1060.7	0.955	0.992	109.9
2	20.88	21.13	0.25	1223.1	1205.7	0.943	0.996	108.8
2	10.17	10.40	0.11	1402.4	1384.6	0.927	0.999	107.4
2	5.01	5.15	0.05	1482.4	1472.5	0.919	1.000	106.6
2	2.49	2.57	0.02	1520.3	1515.9	0.914	1.000	106.3
2	0.00	0.00	0.00	1558.9	1558.9	0.910	1.000	105.9

NRTL Equation Parameters used to Fit Measurements
$\tau_{12} = -0.0445$
$\tau_{21} = -0.0520$
$a = -1.00$

RMS Deviation = 0.98%

Table 10. Vapor-Liquid Equilibrium Measurements on the 1,2-Propanediol (PG)/1-Methoxy-2-propanol (PGME) System at 75°C

Run Num	Mole Percent PG Charge	Liquid	Vapor	Pressure, kPa Meas	Calc	Activity Coefficient PG	PGME	Relative Volatility PGME/PG
1	100.00	100.00	100.00	0.775	0.775	1.000	1.435	36.34
1	97.52	97.52	52.45	1.43	1.44	1.000	1.407	35.65
1	95.19	95.20	36.17	2.06	2.04	1.001	1.383	35.00
1	88.73	88.74	19.14	3.62	3.62	1.005	1.321	33.30
1	80.22	80.24	11.50	5.52	5.50	1.015	1.252	31.24
1	70.01	70.02	7.46	7.52	7.55	1.034	1.184	28.99
2	61.39	61.40	5.52	9.24	9.16	1.056	1.137	27.25
1	60.25	60.26	5.31	9.28	9.36	1.059	1.131	27.03
2	50.72	50.75	3.91	11.11	11.03	1.091	1.090	25.30
1	50.32	50.33	3.86	10.89	11.10	1.093	1.089	25.22
2	40.97	41.00	2.85	12.76	12.68	1.131	1.058	23.66
2	30.40	30.43	1.95	14.54	14.45	1.184	1.031	22.03
2	20.50	20.53	1.24	16.19	16.13	1.243	1.014	20.63
2	10.15	10.17	0.58	17.98	17.94	1.317	1.003	19.27
2	5.35	5.36	0.30	18.83	18.81	1.356	1.001	18.68
2	2.61	2.61	0.15	19.32	19.32	1.379	1.000	18.35
2	0.00	0.00	0.00	19.81	19.81	1.402	1.000	18.04

NRTL Equation Parameters used to Fit Measurements
τ_{12} = 0.1286
τ_{21} = 0.1919
α = -1.00

RMS Deviation = 0.69%

Table 11. Vapor-Liquid Equilibrium Measurements on the 1,2-Propanediol (PG)/1-Methoxy-2-propanol (PGME) System at 200°C

Run Num	Mole Percent PG Charge	Liquid	Vapor	Pressure, kPa Meas	Calc	Activity Coefficient PG	PGME	Relative Volatility PGME/PG
1	100.00	100.00	100.00	8.23	8.23	1.000	1.405	16.59
1	97.38	97.48	70.27	11.28	11.43	1.000	1.387	16.37
1	94.88	95.07	54.42	14.35	14.42	1.001	1.371	16.17
1	89.73	90.06	36.55	20.59	20.43	1.003	1.337	15.73
1	79.68	80.17	21.39	31.65	31.46	1.011	1.274	14.86
1	69.60	70.14	14.40	41.76	41.67	1.027	1.217	13.97
1	59.59	60.08	10.34	51.06	51.06	1.051	1.165	13.05
2	54.24	54.45	8.71	55.58	56.01	1.069	1.138	12.53
1	49.31	49.70	7.55	59.91	60.02	1.088	1.117	12.09
2	42.69	42.94	6.16	65.36	65.56	1.119	1.090	11.46
1	39.56	39.83	5.59	68.01	68.05	1.136	1.079	11.17
2	31.36	31.62	4.26	74.34	74.49	1.190	1.051	10.40
2	21.16	21.39	2.80	82.83	82.42	1.278	1.025	9.433
2	10.32	10.46	1.37	91.96	91.09	1.408	1.006	8.403
2	5.07	5.14	0.68	96.49	95.51	1.489	1.002	7.905
2	2.50	2.54	0.34	98.82	97.75	1.534	1.000	7.662
2	0.00	0.00	0.00	99.98	99.98	1.582	1.000	7.426

NRTL Equation Parameters used to Fit Measurements
τ_{12} = 0.3182
τ_{21} = 0.0214
α = -1.00

RMS Deviation = 0.65%

Table 12. Vapor-Liquid Equilibrium Measurements on the 1,2-Propanediol (PG)/Dipropylene Glycol Monomethyl Ether (DPGME) System at 100°C

Run Num	Mole Percent DPGME Charge	Liquid	Vapor	Pressure, kPa Meas	Calc	Activity Coefficient DPGME	PG	Relative Volatility PG/DPGME
1	100.00	100.00	100.00	3.97	3.97	1.000	1.774	1.421
1	97.19	97.19	96.14	4.00	4.02	1.000	1.735	1.389
1	94.42	94.42	92.58	4.02	4.06	1.001	1.697	1.358
1	89.23	89.24	86.46	4.09	4.12	1.005	1.629	1.299
1	79.93	79.94	76.95	4.16	4.20	1.018	1.517	1.194
1	69.90	69.90	68.25	4.20	4.25	1.044	1.408	1.080
2	60.52	60.52	61.12	4.30	4.26	1.082	1.317	0.975
1	60.03	60.03	60.76	4.26	4.26	1.084	1.313	0.970
2	50.64	50.64	54.24	4.30	4.23	1.141	1.232	0.865
1	47.19	47.18	51.92	4.26	4.22	1.168	1.206	0.827
2	40.38	40.37	47.37	4.16	4.17	1.232	1.157	0.752
1	38.43	38.43	46.05	4.16	4.16	1.253	1.144	0.731
2	30.27	30.26	40.30	4.07	4.07	1.364	1.094	0.643
2	19.83	19.82	31.69	3.89	3.90	1.571	1.044	0.533
2	9.98	9.97	20.40	3.65	3.65	1.879	1.012	0.432
2	5.02	5.01	12.11	3.42	3.45	2.101	1.003	0.383
2	2.52	2.52	6.72	3.32	3.33	2.238	1.001	0.359
2	0.00	0.00	0.00	3.18	3.18	2.397	1.000	0.334

Wilson Equation Parameters used to Fit Measurements
Λ_{12} = 0.4103
Λ_{21} = 1.0166

RMS Deviation = 0.76%

Table 13. Vapor-Liquid Equilibrium Measurements on the 1,2-Propanediol (PG)/Dipropylene Glycol Monomethyl Ether (DPGME) System at 180°C

Run Num	Mole Percent DPGME Charge	Liquid	Vapor	Pressure, kPa Meas	Calc	Activity Coefficient DPGME	PG	Relative Volatility PG/DPGME
1	100.00	100.00	100.00	73.32	73.32	1.000	1.457	1.598
1	96.87	96.95	95.28	74.52	74.66	1.000	1.435	1.573
1	94.47	94.59	91.84	75.67	75.65	1.001	1.418	1.553
1	89.52	89.71	85.21	77.61	77.55	1.003	1.384	1.512
1	79.84	80.08	73.78	80.79	80.75	1.011	1.319	1.429
1	70.82	71.05	64.55	83.05	83.14	1.026	1.262	1.348
2	60.57	60.62	55.16	85.47	85.21	1.053	1.202	1.251
1	55.27	55.40	50.84	85.83	85.98	1.071	1.173	1.201
2	50.22	50.27	46.76	86.57	86.57	1.093	1.147	1.151
1	49.38	49.46	46.12	86.46	86.64	1.097	1.143	1.143
2	40.30	40.32	39.12	87.41	87.23	1.148	1.101	1.051
2	30.46	30.44	31.56	87.21	87.23	1.227	1.062	0.949
2	20.37	20.31	23.24	86.26	86.42	1.344	1.030	0.841
2	10.45	10.37	13.62	84.56	84.55	1.511	1.009	0.733
2	5.29	5.23	7.54	83.05	82.96	1.627	1.002	0.677
2	2.79	2.75	4.17	82.13	81.99	1.694	1.001	0.650
2	0.00	0.00	0.00	80.71	80.71	1.776	1.000	0.619

Wilson Equation Parameters used to Fit Measurements
Λ_{12} = 0.4856
Λ_{21} = 1.1478

RMS Deviation = 0.14%

Table 14. Vapor-Liquid Equilibrium Measurements on the Acetophenone (ACPN)/Propene (C_3) System at 25°C

Run Num	Mole Percent ACPN Charge	Liquid	Vapor	Pressure, kPa Meas	Calc	Activity Coefficient ACPN	C_3	Relative Volatility C_3/ACPN
1	100.00	100.00	100.00	0.037	0.037	1.000	2.967	75098
1	93.68	94.93	0.03	134.45	137.07	1.002	2.799	68843
1	89.50	91.41	0.02	225.39	226.53	1.005	2.688	64645
1	81.57	84.34	0.01	398.52	391.84	1.016	2.478	56574
1	80.25	83.11	0.01	421.27	418.57	1.019	2.443	55222
1	70.77	73.92	0.01	604.87	599.82	1.049	2.195	45565
1	61.20	64.03	0.00	764.83	757.11	1.106	1.953	36127
2	57.41	58.47	0.00	823.23	828.43	1.154	1.827	31267
1	50.97	53.01	0.00	890.52	886.74	1.216	1.710	26817
2	44.52	46.49	0.00	934.10	941.85	1.319	1.579	21934
1	40.27	41.35	0.00	980.09	974.94	1.430	1.482	18424
2	36.68	38.32	0.00	981.47	990.56	1.513	1.427	16500
2	31.56	32.61	0.00	1007.0	1013.4	1.720	1.331	13148
1	28.95	29.19	0.00	1035.4	1023.7	1.887	1.277	11311
2	23.00	24.09	0.00	1036.8	1036.0	2.227	1.203	8811
2	15.39	16.34	0.00	1057.0	1053.2	3.112	1.106	5515
2	8.90	9.61	0.00	1081.8	1076.8	4.665	1.042	3069
2	3.25	3.57	0.00	1124.2	1119.0	7.649	1.007	1651
2	0.00	0.00	0.00	1162.5	1162.5	11.103	1.000	1092

NRTL Equation Parameters used to Fit Measurements
τ_{12} = 0.9085
τ_{21} = 0.1536
a = -1.00

RMS Deviation = 0.79%

Table 15. Vapor-Liquid Equilibrium Measurements on the Acetophenone (ACPN)/Propene (C_3) System at 60°C

Run Num	Mole Percent ACPN Charge	Liquid	Vapor	Pressure, kPa Meas	Calc	Activity Coefficient ACPN	C_3	Relative Volatility C_3/ACPN
1	100.00	100.00	100.00	0.512	0.512	1.000	2.847	9769
1	93.68	95.49	0.23	221.11	222.06	1.001	2.709	9145
1	89.50	92.31	0.14	371.97	372.37	1.003	2.616	8712
1	81.57	85.83	0.08	668.79	664.60	1.012	2.434	7847
1	80.25	84.68	0.07	713.74	713.96	1.015	2.404	7697
1	70.77	75.95	0.05	1076.3	1068.2	1.039	2.180	6577
1	61.20	66.15	0.04	1425.7	1412.5	1.088	1.949	5384
2	57.41	59.12	0.03	1609.9	1621.2	1.142	1.797	4573
1	50.97	54.71	0.03	1745.8	1734.8	1.188	1.706	4087
2	44.52	48.25	0.03	1854.7	1875.8	1.277	1.581	3412
1	40.27	42.16	0.03	1995.3	1981.8	1.395	1.470	2817
2	36.68	39.90	0.03	2003.6	2014.6	1.450	1.431	2609
2	31.56	33.52	0.02	2088.4	2090.8	1.652	1.327	2056
1	28.95	29.08	0.02	2155.3	2131.7	1.851	1.260	1705
2	23.00	25.21	0.02	2169.1	2161.7	2.083	1.206	1421
2	15.39	17.47	0.02	2237.3	2218.0	2.822	1.112	916.0
2	8.90	10.57	0.02	2298.0	2287.7	4.105	1.046	527.4
2	3.25	4.08	0.02	2413.8	2411.1	6.650	1.008	200.1
2	0.00	0.00	0.00	2551.7	2551.7	9.849	1.000	84.72

NRTL Equation Parameters used to Fit Measurements
τ_{12} = 0.8858
τ_{21} = 0.1395
a = -1.00

RMS Deviation = 0.52%

Table 16. Vapor-Liquid Equilibrium Measurements on the Dimethyl Disulfide (DMDS)/Methanethiol (MM) System at 0°C

Run Num	Mole Percent DMDS Charge	Liquid	Vapor	Pressure, kPa Meas	Calc	Activity Coefficient DMDS	MM	Relative Volatility MM/DMDS
1	100.00	100.00	100.00	0.985	0.985	1.000	0.971	77.30
1	97.12	97.15	30.56	3.15	3.14	1.000	0.975	77.53
1	95.29	95.34	20.84	4.48	4.51	1.000	0.977	77.67
1	91.47	91.55	12.19	7.42	7.42	1.000	0.981	77.98
1	76.53	76.69	3.99	19.14	19.05	0.997	0.997	79.15
1	65.93	66.11	2.38	27.56	27.55	0.993	1.006	79.93
1	56.63	56.81	1.61	34.73	35.10	0.989	1.011	80.50
2	45.06	45.27	1.01	44.91	44.50	0.986	1.016	80.91
1	43.98	44.12	0.97	45.17	45.44	0.985	1.016	80.92
2	33.17	33.43	0.62	54.18	54.05	0.986	1.016	80.68
2	21.55	21.80	0.35	63.15	63.20	0.997	1.012	79.18
2	9.91	10.08	0.15	72.30	72.16	1.039	1.004	75.20
2	4.80	4.89	0.07	76.33	76.11	1.079	1.001	72.10
2	2.56	2.61	0.04	77.77	77.88	1.104	1.000	70.38
2	0.00	0.00	0.00	79.94	79.94	1.140	1.000	68.07

Wilson Equation Parameters used to Fit Measurements
Λ_{12} = 0.2992
Λ_{21} = 2.0764

RMS Deviation = 0.49%

Table 17. Vapor-Liquid Equilibrium Measurements on the Dimethyl Disulfide (DMDS)/Methanethiol (MM) System at 100°C

Run Num	Mole Percent DMDS Charge	Liquid	Vapor	Pressure, kPa Meas	Calc	Activity Coefficient DMDS	MM	Relative Volatility MM/DMDS
1	100.00	100.00	100.00	80.67	80.67	1.000	0.986	14.90
1	97.12	97.28	70.63	113.42	112.15	1.000	0.989	14.88
1	95.29	95.54	59.04	133.07	132.51	1.000	0.990	14.86
1	91.47	91.88	43.30	175.13	175.83	1.000	0.994	14.82
1	76.53	77.31	18.89	356.46	355.22	0.998	1.005	14.63
1	65.93	66.73	12.18	488.15	492.24	0.995	1.012	14.46
1	56.63	57.32	8.60	616.39	618.30	0.993	1.015	14.27
2	45.06	45.90	5.72	779.10	775.96	0.991	1.017	13.97
1	43.98	44.38	5.42	794.96	797.36	0.991	1.017	13.92
2	33.17	34.33	3.71	941.82	939.62	0.993	1.016	13.56
2	21.55	22.77	2.22	1105.2	1105.7	1.006	1.011	12.97
2	9.91	10.76	0.99	1290.0	1280.7	1.044	1.004	12.03
2	4.80	5.27	0.48	1374.8	1362.8	1.079	1.001	11.44
2	2.56	2.83	0.26	1408.6	1400.5	1.101	1.000	11.13
2	0.00	0.00	0.00	1445.1	1445.1	1.132	1.000	10.73

Wilson Equation Parameters used to Fit Measurements
Λ_{12} = 0.3322
Λ_{21} = 1.9776

RMS Deviation = 0.54%

ISOTHERMAL VAPOR-LIQUID EQUILIBRIUM OF PROPYLENE OXIDE-ACETOPHENONE SYSTEM

Chen Zhongxiu, Tang Jianhua, Wang Lihua and Wu Zhaoli ■ Department of Chemical Engineering,
Zhejiang University,
Hangzhou, P.R. China

The isothermal p—x data of propylene oxide (1) and acetophenone (2) system were determined at 27 and 32°C using a static p—x method. The experimental data were employed to compute equilibrium vapor compositions y by means of a direct method.

apor — Liquid equilibrium data of propylene oxide(1)—acetophenone(2) has not been found in the literature. The difference between the boiling points of the two components is 167.7°C. In other words, the relative volatility is very large. For such a system, it is difficult to determine the concentration of each phase accurately. This work intended to determine the vapor—liquid equilibrium data of this system to meet the need of process design and performance evaluation.

EXPERIMENTAL APPARATUS

The scheme of the apparatus for vapor—liquid equilibrium measurement is shown in Figure 1. At constant temperature the total pressure p and the composition of liquid phase x can be measured by the apparatus [1]. The equilibrium cell 8 was immersed in a thermostat 19. The temperature was controlled within $\pm$ 0.01°C. The back pressure was controlled by a pressure regulator system within $\pm$ 6.7Pa. The total pressure at equilibrium was measured by the nullmanoneter 9 and the U-manometer 22. A sampling chamber 11 with a water jacket was mounted at the top of the equilibrium cell. By raising the mercury level in the equilibrium cell the liquid phase at equilibrium condition can be introduced into the sampling cell. Thus the sample could be taken for determination of liquid composition x. A vacuum pump 21 was used for evacuating the system.

EXPERIMENTAL PROCEDURE

The procedure of the measurement was as follows. The binary solution from the sampling chamber was introduced into the equilibrium cell 8 by opening valve 3. Then the noncondensable gas was evacuated. Degassing was considered to be completed if the vapor phase disappeared by closing valve 3 and raising the mercury level in equilibrium cell.

The thermostat was controlled at the desired temperature for one hour. By raising the mercury level, the vapor phase in the equilibrium cell was expelled until there remains only a tiny vapor bulb. The mercury levels in the equilibrium cell and the nullmanometer were adjusted to about the same level by changing the back pressure. Then the mercury levels in the equilibrium cell and the nullmanometer, the solution level in the equilibrium cell and the pressure difference in the U-manometer were measured with a cathetometer. Thus the total pressure at equilibrium condition could be measured.

The sampling chamber 11 was washed several times with the light component (propylene oxide), then it was evacuated. The liquid in equilibrium cell was raised into the sampling chamber by raising the mercury level, then a sample was taken for determination of x. Thus, a p-x data point at definite temperature was obtained. By changing the composition of the binary solution and repeating the above mentioned procedure, a series of plots of the p-x curve were determined.

The composition of the equilibrium liquid was determined by gas chromatography. The column was packed by 101 white support produced by Shanghai Chemical Reagent Factory and was coated with methyl silicone-30. Although the difference between the refractive index of propylene oxide and acetophenone is significant, the refractive index method for determination of composition failed because propylene oxide is very hygroscopic and evaporates rapidly.

Our experimental method was tested by measuring the vapor—liquid equilibrium of hexane—benzene at 50°C and comparing with the data in literature. The results are listed in Table 1 and plotted in Figure 2. It can be seen that the data of this work agree well with the data in the literature [2].

The propylene oxide used was a product of ARCO, and acetophenone came from the Shanghai Reagent Company. Their physical properties are listed in Table 2.

RESULTS AND DATA CORRELATION

The experimentally determined vapor pressures of propylene oxide, acetophenone, and their mixtures are listed in Table 3. These results were correlated by direct method [9]. The basic differential equations are derived from Gibbs—Duhem equation and can be solved easily by iterative method.

For vapor—liquid equilibrium of an n-component system the Gibbs—Duhem equations of liquid phase can be written as

$$\sum_{i=1}^{n} x_i d\mu_i = -S^L dT + V^L dp \qquad (1)$$

In view of the definitions of $\hat{f}_i^L$ and S^L this becomes

$$\sum_{i=1}^{n} x_i d\ln \hat{f}_i^L = \frac{H^{id} - H^L}{RT^2} dT + \frac{V^L}{RT} dp \qquad (2)$$

using equilibrium property,

$$\hat{f}_i^L = \hat{f}_i^V = p y_i \hat{\varphi}_i \qquad (3)$$

we obtain

$$\sum_{i=1}^{n-1}(\frac{x_i}{y_i}-\frac{x_n}{y_n})(\frac{\partial y_i}{\partial x_j})_{x[j,n]} + \sum_{i=1}^{n} x_i (\frac{\partial \ln \hat{\varphi}_i}{\partial x_j})_{x[j,n]}$$
$$-\frac{H^{id}-H^L}{RT^2}(\frac{\partial T}{\partial x_i})_{x[j,n]} + (\frac{1}{p}-\frac{V^L}{RT})(\frac{\partial p}{\partial x_j})_{x[j,n]}$$
$$= 0 \qquad (4)$$

At constant temperature this reduces to

$$\sum_{i=1}^{n-1}(\frac{x_i}{y_i}-\frac{x_n}{y_n})(\frac{\partial y_i}{\partial x_j})_{x[j,n]} + \sum_{i=1}^{n} x_i (\frac{\partial \ln \hat{\varphi}_i}{\partial x_j})_{x[j,n]}$$
$$+ (\frac{1}{p}-\frac{\sum_{i=1}^{n} x_i V_i^L}{RT})(\frac{\partial p}{\partial x_j})_{x[j,n]} = 0 \qquad (5)$$

In deriving this result we have used the relations

$$V^L = \sum_{i=1}^{n} x_i V_i^L + V^E \doteq \sum_{i=1}^{n} x_i V_i^L \qquad (6)$$

For binary systems at constant temperature we obtain

$$F \equiv (\frac{x_1}{y_1}-\frac{x_2}{y_2})\frac{dy_1}{dx_1} + x_1 \frac{d\ln \hat{\varphi}_1}{dx_1} + x_2 \frac{d\ln \hat{\varphi}_2}{dx_1}$$
$$+ (\frac{1}{p}-\frac{x_1 V_1^L + x_2 V_2^L}{RT})\frac{dp}{dx_1} = 0 \qquad (7)$$

where F denotes the left—hand sum of equation (5).

Equations(7) can be approximated by its finite difference representation in the following manner. A liquid independent composition coordinates are regarded as being discretized with lattice interval Δx, so that $x_1 = k\Delta x$ for intergral k

We obtain

$$F[k] = (\frac{k\Delta x}{y_1[k]}-\frac{1-k\Delta x}{1-y_1[k]})$$
$$\times (\frac{y_1[k+1]-y_1[k-1]}{2\Delta x})$$
$$+ k\Delta x \frac{d\ln \hat{\varphi}_1}{dx_1}[k]$$
$$+ (1-k\Delta x)\frac{d\ln \hat{\varphi}_2}{dx_1}[k]$$
$$+ (\frac{1}{p[k]}-\frac{k\Delta x V_1^L + (1-k\Delta x)V_2^L}{RT})$$
$$\times \frac{dp}{dx_1}[k] \qquad (8)$$

and start with the initial approximation for assuming ideal vapor phase behavior.

Assume that one has available the rth estimates of $F[k]^r$ at the lattice points. We have the formula

$$0 = F[k]^r + \frac{\partial F[k]}{\partial y_1[k-1]}\delta y_1[k-1]$$
$$+ \frac{\partial F[k]}{\partial y_1[k]}\delta y_1[k] + \frac{\partial F[k]}{\partial y_1[k+1]}$$
$$\times \delta y_1[k+1] \qquad (9)$$

If equation(9) is then repeated for all lattice points, there results, in matrix—vector form,

$$\vec{F}_y \vec{\delta y}_1 = -\vec{F} \qquad (10)$$

Where F_y is tridiagonal matrix.

Equation(9) provides a linear set of equations in the correction terms δy_1 that can be solved by relaxation techniques. Choosing the appropriate factor py we obtain

$$\vec{y}_1^{r+1} = \vec{y}_1^r + \vec{\delta y}_1 \times py \qquad (11)$$

Substituting y_1^{r+1} into equation(8), we repeat the calculation until convergence to permitting deviation is obtained. The solution to the problem lies in choosing the appropriate value for factor py. A convenient method to choose the one is to specify that

$$RMSF^{(r+1)} < RMSF^{(r)} \quad (12)$$

where $RMSF = \sqrt{\sum_k F[k]^2}$

The value of minimal RMSF is also chosen as a criterion for ending iterations.

To evaluate $\frac{d\hat{\varphi}}{dx_1}$ and $\frac{dp}{dx_1}$, we use spline approximation for $\hat{\varphi}-X_1$ and $p-x_1$ data, and then differentiate the splines. Gas phase fugacity coefficients were calculated from the virial equation truncated after the second term. The second virial coefficients were estimated from correlation equation of Tsonopoulos [10].

The results of computation are listed in Table 3. It is shown that the direct method by Hu Ying [9] can be well used to evaluate equilibrium vapor composition for the propylene oxide—acetophenone system.

ACKNOWLEDGMENT

The authors wish to acknowledge Mr. Qin Zehang and Shen Xinhua for taking part in the experimental measurement and would like to express appreciation to the financial support of DIPPR of AIChE.

NOTATION

F	function, defined by eq. (7)
F	function vector
F_y	partial derivative matrix
f	fugacity
H	molar enthalpy
k	lattice pint index
n_0^{20}	refractive index at 20℃
p	total pressure
py	scale factor
p_{VP}	vapor pressure
R	universal gas constant
RMS	sum of squared deviations
S	molar entropy
T	temeprature K
T_b	normal boiling point
T_f	normal freezing point
V	molar volume
x	mole fraction of liquid
y	mole fraction of vapor
δy	modified difference of y
μ	chemical potential
φ	fugacity coefficient

Superscripts

L	liquid phase
V	vapor phase

Subscripts

i,j	component i and j
E	experimental value
C	calculated value

LITERATURE CITED

1. He Lizhong, Wu Zhaoli, Chen Zhongxiu and Hou Yujun, Chem. Eng., (China) 6, 48(1984).

2. Ida P. C. Li, Yiu-Wah Wong, Shinn-Der Chang, and Benjamin C.-Y. Lu, J. Chem. Eng. Data, 4, 492(1972).

3. Reid, R. C. Sherwood, T. K. and Prausnitz. J. M., The Properties of Gases and Liquid, 3rd, ed., McGraw — Hill Book Company, New York, 641(1977).

4. Mellan, I., Industrial Solvents, 2nd ed., Reinhold Publishing Corp., New York, N.Y. (1950).

5. A. Weissberger: Organic Solvents, 3rd ed., p. 249, Wiley (1970).

6. Bott T. R., Sadler H. N. 1., J. Chem. Eng. Data, 11, 25(1966).

7. D. R. Stull., Ind. Eng. Chem., 39, 517 (1947).

8. J. Livingston et al., J. Amer. Chem. Soc., 48, 881(1924).

9. Hu Ying and Ying Xu-gen, J. Chem. Ind, and Eng., (China), 1, 28(1980).

10. Tsonopoulos, C., AIChE J., 20, 263 (1974).

Table 1. Experimental equilibrium data for n-hexane(1) —benzene(2) system at 50 °C

x_1	0.086	0.127	0.310	0.375	0.615	0.810
p, kPa	41.413	42.349	47.513	48.912	52.255	53.759

Table 2. Physical properties of reagents used

Reagents		Propylene oxide		Acetophenone	
		E$_{XPTL.}$	L$_{IT.}$	E$_{XPT L.}$	L$_{IT.}$
T_b, °C			34.3[3]		202.0[5]
T_f, °C			−112[3]		19.62[8]
n_0^{20}		1.3633	1.3657[4]	1.5310	1.53631[5]
p_{VP} (kPa)	25 °C		70.605[6]		0.0493[7]
	27 °C	77.876	76.329[6]	0.0483	
	32 °C	93.663	92.189[6]	0.0681	

Table 3. Experimental and calculated results for propylene oxide(1) and acetophenone (2) system

	t=27℃			t=32℃	
X_{IE}	Y_{IC}	p(kPa)	X_{IE}	Y_{IC}	p(kPa)
0.0000	0.0000	0.048	0.0000	0.0000	0.068
0.0264	0.9759	1.942	0.0242	0.9766	3.149
0.0795	0.9931	6.954	0.0702	0.9906	7.789
0.1468	0.9967	12.792	0.0863	0.9922	9.183
0.1621	0.9970	13.943	0.1287	0.9949	17.386
0.2486	0.9982	20.833	0.1750	0.9964	18.262
0.2863	0.9984	23.450	0.2732	0.9978	27.229
0.3092	0.9987	25.342	0.3421	0.9984	34.735
0.3930	0.9991	33.038	0.3933	0.9987	38.592
0.4249	0.9992	35.734	0.4061	0.9988	40.185
0.4735	0.9993	37.614	0.5203	0.9993	50.774
0.5352	0.9994	43.525	0.6319	0.9996	61.198
0.6219	0.9995	48.631	0.6899	0.9996	65.629
0.6932	0.9996	53.987	0.7798	0.9997	74.237
0.7945	0.9997	62.175	0.9497	0.9999	88.360
0.8702	0.9998	67.741	1.0000	1.0000	93.663
0.9069	0.9999	70.732			
0.9586	0.9999	74.029			
1.0000	1.0000	77.876			

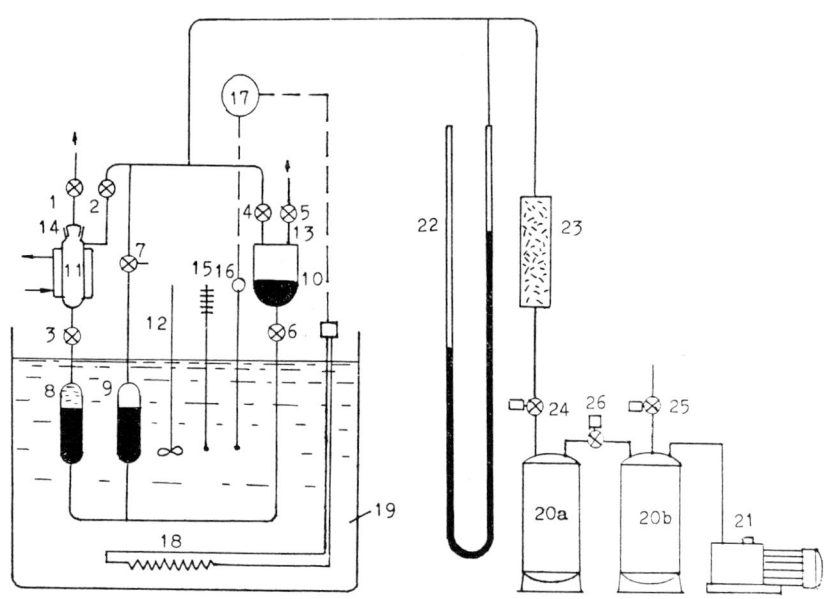

Figure 1. Schematic diagram of apparatus

1 to 6	—	two—way cocks;	7	— three—way cocks;
8	—	equilibrium cell;	9	— nullmanometer;
10	—	mercury reservoir;	11	— sampling chamber;
12	—	stirrer;	13	— neck for adding mercury;
14	—	neck for adding sample;	15	— precise thermometer;
16,17	—	temperature controller;	18	— heater;
19	—	water bath;	20$_{a,b}$	— buffers;
21	—	vacuum pump;	22	— U-manometer;
23	—	absorb vessel;	24,25,26	— magnetic valuves

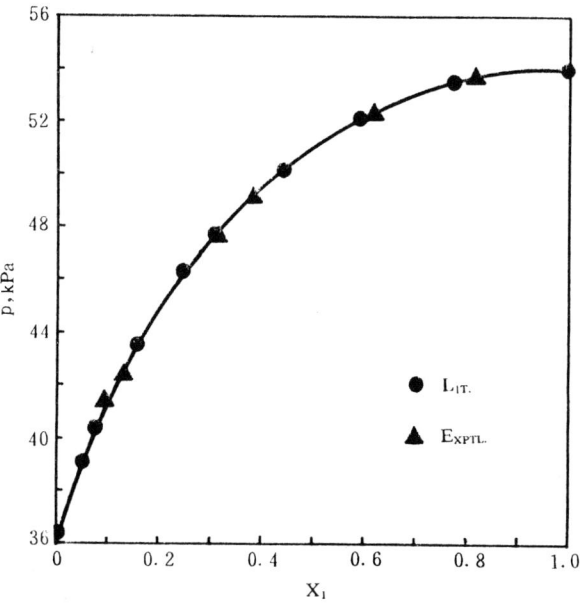

Figure 2. p-x diagram for the system n-hexane(1)—benzene(2) at 50℃

VAPOR-LIQUID EQUILIBRIUM IN BINARIES OF SULFOLANE WITH ORTHO-XYLENE, OCTANE AND ETHYL-CYCLOHEXANE: REPORT ON SUBPROJECT DIPPR 805NSF/88/B

David Zudkevitch and Deepak Shukla ■ Department of Chemical Engineering,
Columbia University
New York, New York 10027

Experimental (X, Y, t, and P) data on three binary mixtures of sulfolane and three hydrocarbons were obtained in an isothermal circulating glass still. Each system was studied at two temperatures under subatmospheric pressures. The binary system, ortho-xylene (OXY) + Sulfolane (SUL), was studied at 110 C and 142 C through the entire composition range. The system of n-octane (OCT) + sulfolane (SUL) was studied at 99 C and 123 C and that of ethyl cyclohexane (ETC) + sulfolane (SUL) was studied at 105 C and 132 C. Due to the very limited miscibilities of the components in the systems OCT/SUL and ETC/SUL, only a few data points were taken. Analyses of the data indicated that the binary OXY/SUL exhibits moderate deviations from ideal mixing in the liquid phase under the conditions, while the systems OCT/SUL and ETC/SUL exhibit strong deviations from ideal mixing and very limited mutual miscibility. In all systems, high relative volatilities were observed between the hydrocarbons and the high boiling sulfolane. Consequently, difficulties were encountered in data measurements and analyses; the systems are very sensitive to minute variations in temperature and pressure.

INTRODUCTION

Incentive for studying binaries of sulfolane and hydrocarbons is provided by the need for data for designing commercial systems. Another objective has been the desire to reconcile differences and augment available data on sulfolane/hydrocarbon systems (1,3,5,6,9,10,13,17,18)

Although sulfolane is an important solvent commonly used in extraction of aromatics from paraffins and naphthenes, especially in treatment of lubricant oil fractions, few data are in the literature on the VLE of its systems. Some of the data have been considered questionable. The main reason for scarcity of data is the high boiling point of sulfolane, 285 C and the basic fact that sulfolane and paraffins and naphthenes are almost insoluble, hence large activity coefficients.

Experimental determination of VLE data on the binaries of sulfolane and light hydrocarbons, up to ten carbon atoms, has been difficult because of the low solubility and the high relative volatilities ($\alpha > 5000:1$). In all cases the vapor is virtually 99+% light hydrocarbon. Therefore, variations in vapor composition as function of the liquid compositions are only in the third or even fifth decimal place. Moreover, sulfolane's melting point is 27.05 C which further complicates the experimentation.

EXPERIMENTAL APPARATUS

The Columbia-circulating-glass-cell, shown in Figure 1, has been used for measuring VLE data at Columbia University and in other laboratories (20).

This miniature cell was designed for obtaining thermodynamically consistent data while using only small quantities of relatively hard-to-study samples, especially under vacuum. Its operation is as close as possible to isothermal. The liquid is stirred magnetically and the vapor dome is heated by an external coating of film (Trade Name: Instatherm). With three separate automatically-controlled heating circuits, practically uniform temperature is maintained inside the entire cell. This feature differentiates the cell's operation from those of most condensate-circulating cells in which the temperature of the vapor, which is colder than the liquid, is reported (12).

The condensate is allowed to flow back into the still until equilibrium is reached and when a sample is desired, the three-way stopcock is turned and a small sample, e.g., 0.1 ml or less, is allowed to accumulate above the stopcock and a sample is taken with a syringe for GC analysis of the condensate, i.e., vapor. Following sampling the stopcock is turned for total circulation. This procedure minimizes the condensate holdup and, hence, depletion of the liquid of the volatile component.

An additional advantage of this small apparatus is that experimental conditions, e.g., pressure and composition, can be changed quickly.

THEORY AND EVALUATION OF DATA

At equilibrium the chemical potentials of component i in the liquid and the vapor are equal.

$$\mu_i^L = \mu_i^V \quad (1)$$

In terms of fugacity equality of a binary point under low-to-moderate pressures, equation 1 comes:

$$P Y_i \hat{\phi}_i^V = p_i^o \phi_i^o X_i \gamma_i (PE)_i \quad (2)$$

where $(PE)_i$ is the Poynting effect of i; ϕ_i^V and ϕ_i^o are the fugacity coefficients of i in the vapor mixture and as a pure compound respectively.

Per the Gibbs-Duhem equation the validity of the experimental data pair/point is determined from the derivatives, which can only be established by differentiating a series of data points.

$$X_1 \left[\frac{d\ln\gamma_1}{dX_1} \right]_{P,T} = X_2 \left[\frac{d\ln\gamma_2}{dX_2} \right]_{P,T} \quad (3)$$

The derivatives of equation 3 are point properties and therefore should be both isothermal and isobaric. The dual requirement can be satisfied only at a single point; it can't be satisfied for a family of points, e.g. an isotherm of a mixture. Consequently, empirical equations, each representing an integration of equation 3 are used in the analysis and correlation of experimental data.

Theoretically, either isothermal or isobaric data can be analyzed with an integral form of equation 3. However, the heat of mixing, at each point, which is usually unknown, enters the isobaric form of equation 3. Therefore, measuring isothermal data has been preferred.

Per Redlich and Kister (14), integration of the activity coefficient ratios, $\log \gamma_1/\gamma_2 = f(X_1)$ of thermodynamically consistent data should produce the zero sum shown in equation 4.

$$\int_0^1 (\ln\gamma_1 - \ln\gamma_2) dX_1 = 0 = \int_0^1 \ln\frac{\gamma_1}{\gamma_2} dX_1 \quad (4)$$

Redlich and Kister recommended a polynomial, up to the fifth order, for the integrand of equation 4. Frequently, a VLE system is characterized by the highest power required for its correlation.

Experimental VLE data are correlated/reduced for testing the consistency, i.e. validity of assumptions and reliability of measured and recorded data. Per theoretical considerations, that have been discussed in several publications, most prominently by Van Ness (19), only three of the four variables (PTXY) suffice for calculating the activity coefficients. Consequently, only PTX data are measured and recorded by many researchers. The latter approach relies on the assumption that the experimental procedure and instruments are perfect and the consistency is taken for granted. In the authors' opinion, very few experiments are set so that perfect data can be obtained. Therefore, the authors believe that only by analyzing and cross-checking data and predictions, can the quality of the data be assessed. In this study, PTXY data were taken redundantly. Pressures (subatmospheric) were measured by two manometers., absolute and relative, and several thermocouples were used to indicate the temperatures at different parts of the cell.

The Redlich-Kister thermodynamic consistency test of comparing the negative and positive parts of the graphic integral of log γ_1 / γ_2 as function of X_1 can be used in a modified form (8) in evaluating data on partially miscible systems

$$A = 100 \times |A^*|$$

(5)

The variables are defined below.

In a partially miscibile isothermal system, when the excess volume effect is negligible, the test is:

$$A^* = \int_0^{X1} \ln \frac{a_1}{a_2} \, dX1$$

$$+ (X1'' - X1') \ln \left(\frac{a_1}{a_2}\right)^S$$

$$+ \int_{X1''}^1 \ln \frac{a_1}{a_2} \, dX1$$

$$+ \int \frac{V^M}{RT} \left(\frac{dp}{dx_1}\right)_T dx_1$$

(6)

Gilmont (in 1947) proposed a correlation which is essentially equivalent to the Redlich-Kister equation. A second order truncated form of the equation was used by Gilmont, Zudkevitch, and Othmer (7), and is shown in equation 7.

$$\log \frac{\gamma_1}{\gamma_2} = g'(X_2 - X_1) + g'' \left[(X_2 - X_1)^2 - \frac{1}{12}\right]$$

(7)

Equation 7 has been found adequate for describing most systems encountered. In analyzing the experimental data of this work, the correlation parameters, g' and g" were established by the method of weighted least squares. Since the computer program's logic is limited, the log γ_1 / γ_2 vs. X_1 were also plotted and viewed for the final decision on correlation parameters and whether additional data and repeats were needed.

The coefficients A_{12}, B_{12}, A_{21}, and B_{21} of the three suffix Margules equation,

$$\log \gamma_1 = A_{12}^M X_2^2 + B_{12}^M X_2^3 \tag{8a}$$

and

$$\log \gamma_2 = A_{21}^M X_1^2 + B_{21}^M X_1^3 \tag{8b}$$

are related to the coefficients g'_{12} and g''_{12} by

$$A_{12}^M = 0.5(g' - g'') \tag{9a}$$

$$B_{12}^M = \frac{2}{3} g'' \tag{9b}$$

$$A_{21}^M = 0.5(g' + g'') \tag{9c}$$

and

$$B_{21}^M = -\frac{2}{3} g'' \tag{9d}$$

Similarly, the coefficients of the van Laar equation, per Carlson and Colburn (2), are established from $g'12$ and $g''12$ by:

$$A_{12}^M = 0.5\left(g' + \frac{1}{3} g''\right) \tag{10a}$$

and

$$A_{21}^M = 0.5\left(g' - \frac{1}{3} g''\right) \tag{10b}$$

As presented in Tables 4, 8, and 12, the vapor compositions and total pressures were calculated with equations 9a and 9d. The validity of the data, assumptions, and the correlations were tested. The deviations in predictions of vapor compositions are presented, for each component i, as:

$$\Delta Y_i = \frac{Y_{icalc.} - Y_{iexp.}}{Y_{iexp.}} \times 100 \tag{11}$$

STUDY: BINARY SYSTEM ORTHO-XYLENE/SULFOLANE

Materials

The Ortho-Xylene of 97+ % purity was purchased from Aldrich. A batch of 1000 ml of the material from Aldrich was batchdistilled in a 30 tray 1" Oldershaw column at 10:1

reflux and a heart-cut of 99.4% was collected for the VLE studies.

Sulfolane of 99+% purity was received from Phillips Petroleum Co. and was redistilled, as above, to produce a heart-cut of 99.9+% purity for vapor pressure and VLE experiments.

Because of the high hygroscopy of sulfolane, the open bottles of sulfolane and samples of binary mixtures that contained sulfolane had to be kept in a dessicator.

The sulfolane which is solid at room temperature was melted, in a hot-water bath, prior to each experiment.

Experimental and Analytical Procedure

The vapor pressures of the pure components were measured at the nominal temperatures of the isothermal VLE studies. These data, are presented in Table 1, which is a summary of relevant physical properties and with the VLE data in Tables 2. In the VLE studies, each liquid mixture was prepared by adding a predetermined quantity of Ortho-Xylene to the cell's contents to reach the volume of 40 ml liquid of a desired composition in the cell. The liquid in the cell was then brought to 110 C, or about, the two temperature controllers were set so that the vapor leaving the cell could be maintained at either the same or at slightly (about 0.5 C) higher temperature than that of the liquid. Fine tuning of the temperatures was done by manually controlling the system's pressure. The powerstats were set to achieve a rate of vaporization/condensation of 5 to 20 minutes after the temperature and pressure levels were stabilized.

When the system reached equilibrium, all conditions, five thermocouples' and two manometers' readings were recorded. The three-way condensate-stopcock was turned to allow about 0.25 ml of liquid to accumulate. Following that, the vacuum in the system was broken; the tent valve was opened and the vacuum pump was turned off to prevent vibrations and fluctuations.

About a minute later, the stopcocks of the sampling valves were turned so that the syringe could be introduced through them into the condensate receiver and the bottom of the cell. Samples of the liquid and the condensate were withdrawn, one at a time, with a syringe through the sampling septa, for injection in the GC. At least two samples of each phase were analyzed to insure reproducitility. When a sample of sulfolane-rich liquid was taken, the syringe was kept warm lest the sample froze before injection.

The temperature controllers' settings were then changed to the other temperature, i.e. 142 C, and the pressure controlling procedure was repeated to reach equilibrium at the new temperature.

When it was deemed necessary, predetermined amounts of liquid were withdrawn from the cell to be replaced by predetermined mixtures to make the next desired solutions.

Analyses were carried out by a thermal conductivity GC in Hewlett Packard TC 5830 apparatus; a 6 ft x 1/8" OD stainless steel column filled with the Supelco SE-54 silicone adsorbant on Supelport 100/120 mesh 3% loading was used. Small (0.5 microliter) samples of the liquid were injected. However, since the condensate samples contained so little sulfolane, large doses, 2 microliter, were injected. The analytical operation was temperature programmed through mid-course changes in slope-sensitivity and attenuation were made for measuring the area % of the minute concentration of the sulfolane.

Due to the low solubilities and high relative volatilities, the vapor samples contained very little sulfolane and the GC had to be run at very high sensitivity. The area % which were obtained had significant scatter and the injections had to be repeated in order to obtain as much data as needed for rounding off and averaging.

For verification, several samples were also analyzed, in another laboratory. Those samples were injected into Hewlet-Packard 5840 GC; the capillary was filled with DD-624 (Cyanopropyl phenyl derivative) manufactured by J & W Scientific Co.

The analytical data, in area %, were converted to concentration in mole %. A practically-linear relationships between mole % and area % were established from analyses of standard mixtures.

Results, Analysis, and Correlation of Data

Fourteen PTXY data (X,Y) pairs, were obtained at two isotherms, 110 C (seven points) and 142 C (seven points). For consistency, the vapor pressures of the two components were also measured at the nominal temperatures of the experiments. The vapor pressure data are presented in Table 1 and also, together with the VLE data, in Table 2.

Review of the data in Table 2, and the correlation parameters and comparisons in Tables 3, 4 and 5, and Figures 2, 3, 4 and 5, led to the conclusion that significant, though not very large positive deviations from ideal mixing in the liquid phase within the temperature range of the study. No azeotropes or evidence of hygrodichotomy were observed.

Samples left in the cell for several hours at the high temperature, 142 C, showed some discoloration. However, analyses did not show significant changes in composition.

STUDY: BINARY OF: N-OCTANE (OCT)/SULFOLANE (SUL)

Materials

The pure compounds, n-Octane of 99.9+ % purity was purchased from Aldrich. As in the study of the binary OXYL/SUL, the Sulfolane was taken from the batch generated by redistillation of the material which had been donated by Phillips Petroleum Co.

Experimental and Analytical Procedures

Due to the high relative volatility of OCT/SUL and the high melting point of sulfolane, it was necessary to insure that the temperature of the condenser's walls, i.e. the cooling water in the glass condensers, exceeded 35 C. Otherwise, and it was observed, minute quantities of sulfolane may coat the condenser.

With the following exception, the experimental procedure which is described above was also followed in studying the OXYL/SUL system. Data were taken at the nominal temperatures of 99 and 123 C. Also, since octane and sulfolane mix only to very limited extent, the data were taken at the extreme compositions, 0.4 to 2.4 mole % sulfolane in octane, and one point in the sulfolane rich side.

Although preliminary tests were made to establish the solubility boundaries and compare them with literature values (6,10), no attempts were made to measure the exact solubility limits.

In order to overcome the compounded problems of high relative volatility and low mutual solubilities, the cell-charges had to be carefully calculated so that the condensate would not contain all the light compound in the charge. For that purpose, the three-way stopcock was kept shut after each run, at a temperature above 35 C, so that the small quantity of condensate at each run, were not allowed to return to the cell while samples of the liquid were analyzed. Only when the analytical results on both phase samples and the calculated material balance were in agreement, could the run be continued for repeated sampling and analysis.

Analyses were carried out in the same GC apparatus and the same packed column which had been used for the system OXY/SUL.

Results, Analysis, and Correlation of Data

Twelve PTXY data (X,Y) pairs, of vapor condensate and liquid, were measured at two temperatures, six at around 99 C and six at around 123 C. All the data were presented in Table 6. Also, the vapor pressures of the two components were measured at the same temperatures. These are presented in Tables 1 and 6.

At both temperatures, the system OCT/SUL exhibits strong deviations from ideal mixing and the two liquid phases, each containing about 98 mole % of one of the compounds, could be observed. For that reason, and following preliminary tests and computations, the experiments were carried out with the cell charges so selected that only one liquid phase was in either the equilibrium cell or the condensate receiver.

The experimental data were analyzed and curve-fitted with equation 7, which implies a consistency test. Also the ratios were plotted for the extrapolation to the limits.

Equations 5 and 6 could not be used in testing the thermodynamic consistency of the data on the system OCT/SUL, since only one point was taken in the sulfolane-rich side of each system. Hence graphic integrations for consistency tests of this partially miscible system was not carried out.

The activity coefficients at infinite dilution, thus obtained were compared with the values obtained from regressions. Selected correlation parameters are presented in Table 7.

A comparison of calculated-vs.-experimental values is given in Table 8 and is also summarized in Table 9. In reviewing the comparisons, attention must be paid to the minute variations in the temperatures of the individual experiments and to the extreme conditions, i.e. low concentrations of sulfolane.

In view of these constraints, the data seem reasonably good.

STUDY: BINARY OF: ETHYL CYCLOHEXANE (ETC)/SULFOLANE (SUL)

Materials

The pure compounds, Ethyl cyclohexane of 99+ % purity was purchased from Aldrich. As in the study of the binary OXYL/SUL, and the OCT/SUL, Sulfolane was taken from the batch (heart cut) generated by redistillation of the material which had been donated by Phillips Petroleum Co.

Experimental and Analytical Procedures

Due to the high relative volatility of ETC/SUL and the high melting point of sulfolane, it was necessary insure that the temperature of the condenser's walls, i.e. the cooling water in the glass condensers, exceeded 35 C. Otherwise, and it was observed, minute quantities of sulfolane may coat the condenser.

The experimental procedure followed in the study of the OCT/SUL system, which is described above, was also followed in studying the ETC/SUL system. Data were taken at the nominal temperatures of 105 and 132 C. Also, since ethyl cyclohexane and sulfolane mix only to very limited extent, the data were taken at the extreme compositions, 0.4 to 4.5 mole % sulfolane in ethyl cyclohexane and one point in the sulfolane rich side.

Although preliminary tests were made to establish the solubility boundaries, no attempts were made to measure the exact solubility limits.

In order to overcome the compounded problems of high relative volatility and low mutual solubilities, the cell-charge had to be carefully calculated so that the condensate would not contain all the light compound in the charge. For that purpose, the three-way stopcock was kept shut after each run, and the condenser, the condensate-receiver and the reflux tube were kept at a temperature above 35 C. The small quantity of condensate collected at each run, was not allowed to return to the cell while samples of the liquid were analyzed. Only when the analytical results on both phase samples and the calculated material balance were in agreement, could the run be continued for repeated sampling and analysis.

Analyses were carried out in the same GC apparatus and the same packed column which had been used for the systems OXY/SUL and OCT/SUL.

Results, Analysis, and Correlation of Data

Twelve PTXY data (X,Y) pairs, of vapor condensate and liquid, were measured at two temperatures, six at around 105 C and six at around 132 C. All the data were presented in Table 10. Also, the vapor pressures of the two components were measured at the same temperatures. These are presented in Tables 1, and 10.

At both temperatures, the system ETC/SUL exhibits strong deviations from ideal mixing and two liquid phases, each containing about 96 to 99+ mole % of one of the compounds, could be obtained. For that reason, and following preliminary tests and computations, the experiments were carried out with the cell charges selected so that only one liquid phase was in either the equilibrium cell or the condensate receiver.

The experimental data, though within limited composition ranges, were analyzed and curve-fitted with equation 7, which also implies a thermodynamic consistency test. In addition, the ratios were plotted and the activity coefficients at infinite dilutions were obtained by extrapolations and were reconciled with those from the least-squares regressions.

As in the treatment of the data on the system OCT/SUL, Equations 5 and 6 and the graphic integration could not be used in testing the thermodynamic consistency of the data on the system ETC/SUL, since only one point was taken in the sulfolane-rich side of each.

Selected correlation parameters are presented in Table 11. The comparison of calculated-vs.-experimental values is given in Table 12 and is summarized in Table 13.

As in the case of the system OCT/SUL, a review of the comparisons, must take into account the effects of minute variations in the temperatures of the individual experiments and to the extreme conditions, i.e. low concentrations of sulfolane. In view of these constraints, the data seem reasonably good.

DISCUSSION

The data and the correlations verified conclusions from other studies (1,3,5,6,9,10,13,17,18). The deviations from ideal mixing increase from the mixture Ortho-Xylene/sulfolane, which is completely homogeneous in the liquid phase, they are much larger in the system Ethyl-cyclohexane/Sulfolane and even larger in the system Octane/Sulfolane. As had been expected, the systems of Sulfolane with the nonaromatic hydrocarbons exhibit very limited mutual solubilities.

The large deviations from ideal mixing combined with the very large relative volatilities between eight-carbon compounds and sulfolane result in very limited ranges of composition to study. Consequently, analyses and correlations of the phase behavior of the systems n-octane/sulfolane and Ethyl cyclohexane/sulfolane had to be done with individual data points, five in the hydrocarbon-rich side in each isotherm, which are only slightly different from each other. Even with care and repetitions, the scatter in the data could not be avoided.

Solubility limits predicted by the van Laar equation were compared with values offered by Embry (6). Comparison with the data of Mukhopadhyay and Pathak (10) was not done as it would have required extrapolation in both temperature and molecular size.

ACKNOWLEDGEMENT:

This research was sponsored by the Design Institute for Physical Property Data of the American Institute of Chemical Engineers as Subproject 805/88/B for which the authors are grateful. The assistance of Dr. William Parish and Dr. Dale Embry of Phillips Petroleum Co. who supplied the refined sulfolane and advised on analyzing sulfolane systems is greatly appreciated. The advice on analytical procedures and the assistance of Dr. James M. Hanrahan of Allied Signal Corp. contributed greatly to this study.

NOTATION:

μ^L_i - chemical potential of component i in liquid phase.
μ^V_i - chemical potential of component i in vapor phase.
$(PE)_i$ - Poynting correction to fugacity of component i.
ϕ_i - fugacity coefficient of i in mixture in vapor phase.
ϕ^o_i - fugacity coefficient of pure i in vapor phase.
γ_i - activity coefficient of component i in liquid phase.
g' - correlation parameter of the Gilmont equation.
g" - correlation parameter of the Gilmont equation.
A_{ij} - correlation coefficient for components i and j.
B_{ij} - correlation coefficient for components i and j.
L - denotes property in the liquid phase.
M - denotes coefficient of the Margules equation
ETC - denotes Ethyl Cyclohexane
OCT - denotes n-Octane
SUL - denotes Sulfolane
P - Pressure (torr).
p^o_i - Vapor pressure of pure component i at temperature
T - Temperature (degrees C).
V - denotes property in the vapor phase.
v - denotes coefficient in the van Laar equation.
X_i - mole fraction of component i in liquid phase.

Y_i - mole fraction of component i in vapor phase.

^ - denotes property of component in a mixture

LITERATURE CITED

1. Aschcroft, S.J., "Isothermal vapor-liquid Equilibria for the Systems Toluene-n-heptane, Tolene-propane-2-ol, Toluene-Sulfolane and Propane-2-ol-Sulfolane", Jour. of Chem. and Eng. data, 24, (3) 195 (1979).

2. Carlson, H.C. and Colburn, A.P., "Vapor-Liquid Equilibria of Nonideal Solutions," Ind. & Eng. Chem. 34, No. 5, 581, (May, 1942).

3. Casteel, J.R. and Sears, P.G., "Dielectric Constants, Viscosities and Related Physical Properties of 10 Liquid Sulfoxides and Sulfones at Several Temperatures," Jour. of Chem. Eng. Data, 19, (3), 196, (1974).

4. Daubert, T.E., and Danner, R.P., "Design Institute for Physical Property Data-Project 801," AIChE-Pennsylvania State University (1988).

5. De Fre, R.M. and Verhoeye, L.A., "Phase Equilibrium in Systems Composed of an Aliphatic and an Aromatic and Sulpholane," J. Appl. Chem. Biotechnol, 26, 469-487 (1976)

6. Embry, Dale, Letter to D. Zudkevitch, (July 22, 1988)

7. Gilmont, R., Zudkevitch, D., and Othmer, D.F., "Correlation and Prediction of Binary Vapor-Liquid Equilibria," Ind. & Eng. Chem. 53, No. 3, (March, 1961).

8. Kojima, K., Ochi, K. and Moon, H-M., "Thermodynamic Consistency Test of Vapor-Liquid Equilibrium Data-Methanol/Water, Benzene/Cyclohexane and Ethylmethylketone/Water System", paper presented at the Fourth Codata Symposium on Critical Evaluation of Phase Equilibria in Multicomponent Systems, Gradisca D' Isonzo, Italy, (September 4-6, 1989)

9. Mellan, I., "Industrial Solvents Handbook", Noyes Data Corp., (1970).

10. Mukhopadhyay, M. and Pathak, A.S., Infinite Dilution Activity Coefficients from Ebulliometric Isobaric Bubble Point Composition Data of Hydrocarbons-Sulpholane Systems" Jour. Chem Eng. Data 31, 148-152 (1986)

12. Othmer, D.F., Gilmont, R., and Conti, J.J., "An Adiabatic Equilibrium Still for more Accurate Vapor-Liquid Equilibrium Data", Ind. & Eng. Chem., 52, 625, (July, 1960).

13. Rawat, B.S., Gulati, I.B. and Malik, K.L., "Study of Some Sulphur-Group Solvents for Aromatics Extraction by Gas Chromatography", Jour. Appl. Chem. Biotechnol., 26, p 247-252, (1986)

14. Redlich, O., and Kister, A.T., Ind. & Eng. Chem., 40, 345, (1948).

15. Renon, H., and Prausnitz, J.M., "Local Compositions in Thermodynamic Excess Functions for Liquid Mixtures", **AIChE Jour.**, 14, No. 1, 135, (January, 1968).

16. Scott, D.W., Finke, H.L., Gross, M.E., Guthrie, G.B., and Huffman, H.M., Jour. Am. Chem. Soc. _72_, 2424, (1950).

17. Shell Development Co. "Shell Sulfolane-W," Technical Bulletin SC: 798-87, (1987)

18. Van Aken, A.B., and Broensen, J.M., Selectivity and Solvency Properties of Extraction Solvents and their Mixtures," presented at International Solvent Extraction Conference, Toronto, Canada, Sept. 10-17, (1977)

19. Van Ness, H.C., Byer, S.M., and Gibbs, R.E., "Vapor-Liquid Equilibrium: An Appraisal of Data Reduction Methods," AIChE Jour. **19**, 238 (1973).

20. Wileman, K.P., Oelert, Mullins, S.R., and Manley, D.B., "Experimental vapor-Liquid Equilibria for the Methanol/Dimethysulfide System," Report to the AIChE DIPPR on Project 805/NSF(C)/87, (December 14, 1987). _AIChE Symposium Series No. 271_, **85**, 94 (1989).

21. Wilson, G.M., "Vapor-Liquid Equilibrium XI. A New Expression for the Free Energy of Mixing," Jour. Am. Chem. Soc. _86_, 127, (1964).

22. Zudkevitch, D., Forman, A.L. and Deatherage, W.G., "Vapor-Liquid Equilibrium in Binary Mixtures of Ortho-Dichlorobenzene + N-Methyl Pyrrolidone and Methanol + Dimethyl sulfide," Report on Part of DIIPR Project 805/88/B, (April, 1989). _AIChE Symposium Series No. 279_, **86**, 47 (1990).

Table 1. Properties of Components at Conditions

COMPOUND	O-XYLENE	OCTANE	ET-CYCLH	SULFOLANE
Formula Weight	106.17	114.23	112.22	120.17
Melting Point tF C	-25.	-57.	-111.	27.05
Normal Boiling Point tB C	144.45	125.65	132.	284.95
Specific Gravity @ 20 C, gms/cc	0.870	0.7030	0.7880	1.2606
Data from this study	Vapor Pressure (Torr)			
at 98.8 C	--	336.1	--	1.13
" 104.6 C	--	--	348.6	1.49
" 110.1 C	276.1	--	--	2.36
" 122.7 C	--	701.	--	4.00
" 132.0 C	--	--	764.0	5.95
" 142.0 C	711.	--	--	9.70

Table 2. System: Ortho Xylene/Sulfolane Experimental Vapor-Liquid Equilibrium Data

t C LIQUID	VAPOR	P Torr	X1	Y1	γ_1	γ_2	$\log \gamma_1/\gamma_2$
110.1	110.6	2.36	0.0	0.0	-	1.0000	
110.0	110.6	32.9	0.0367	0.9263	2.9972	1.0666	0.4487
110.6	111.0	101.5	0.1435	0.978 4	2.4982	1.0846	0.3623
110.1	110.6	157.7	0.3023	0.987 7	1.8599	1.178	0.1983
110.1	110.4	182.7	0.3923	0.990 85	1.6657	1.1656	0.1551
110.0	110.6	200.4	0.4679	0.992 20	1.5338	1.2448	0.0907
110.0	110.4	238.4	0.7615	0.994 91	1.1243	2.1559	-0.2828
110.1	110.8	254.0	0.8937	0.997 10	1.0036	2.9348	-0.4547
110.1	110.6	276.1	1.0000	1.000 0	-	-	-
142.0	142.3	9.70	0.0000	0.0000	-	1.0000	-
142.0	142.4	106.0	0.0508 0	0.91210	2.678	1.0072	0.4247
142.0	142.4	232.2	0.1418 3	0.96270	2.218	1.0356	0.3308
141.9	142.5	382.3	0.2834 0	0.97980	1.865	1.1106	0.2251
142.0	142.5	507.7	0.4879 3	0.98760	1.446	1.2615	0.0593
141.9	142.3	577.3	0.7025 0	0.99150	1.150	1.699	-0.1699
141.9	142.4	587.1	0.7495 0	0.99215	1.096	1.8961	-0.2378
142.1	142.6	649.0	0.8936 5	0.99505	1.014	3.0860	-0.4833
142.0	142.6	711.0	1.0000 0	1.00000	1.000	-	-

Table 3. Coefficients for Correlating Activity Coefficients System: Ortho Xylene/Sulfolane

TEMPERATURE	110.0 C	142.0 C
CORRELATION COEFFICIENTS		
g'12	1.0797	1.055
	-0.4351	-0.598
g"12		
γ^{∞}_{12}	2.924	2.838
γ^{∞}_{21}	4.122	4.140
Margules A_{12}	.7574	0.827
A_{21}	-0.2900	-0.399
B_{12}	0.3229	0.228
B_{21}	0.2900	0.399
Van Laar A_{12}	0.4686	0.453
A_{21}	0.6155	0.617
NRTL G_{12}	--	--
G_{21}	--	--
α_{12}	--	--

Table 4. Predicted Compared with Experimental VLE Values System: Ortho Xylene/Sulfolane

tC	X1	Y1 EXP.	Y1 CALC.	Y2 CALC.	P (Torr) EXP.	P (Torr) CALC.
110.0	0.03670	0.92630	0.92492	0.07508	31.6	30.31
	0.14348	0.97840	0.97842	0.02158	101.5	95.31
	0.30228	0.98770	0.98850	0.01145	157.7	156.71
	0.36228	0.99085	0.99063	0.00937	182.7	178.56
	0.46793	0.99220	0.99459	0.00823	200.4	192.30
	0.76150	0.99491	0.99495	0.00505	238.4	230.41
	0.89365	0.99710	0.99709	0.00291	254.0	251.27
142.0	0.05080	0.91210	0.91103	0.08897	106.0	104.19
	0.14183	0.96270	0.96455	0.03545	232.2	239.88
	0.28340	0.97980	0.98054	0.01946	382.3	383.95
	0.48793	0.98760	0.98742	0.01258	507.7	504.75
	0.70250	0.99150	0.99115	0.00885	577.3	578.15
	0.74950	0.99215	0.99198	0.00802	587.1	593.50
	0.89365	0.99505	0.99541	0.00459	649.0	652.94

Table 5. Deviations Between Predicted and Experimental VLE Values System: Ortho Xylene/Sulfolane

X1	DY(1)*	DY(2)*	DELP (Torr)	% DELP
t=110.1 C 0.03670	-0.149	1.876	-1.29	-4.08
0.14348	0.002	-0.101	-6.19	-6.10
0.30228	0.086	-6.875	-0.99	-0.63
0.39228	-0.023	2.449	-4.13	-2.26
0.46793	-0.043	5.462	-8.09	-4.04
0.76150	0.005	-0.882	-7.89	-3.35
0.89365	-0.001	0.346	-2.72	-1.07
t=142.0 C 0.05080	-0.12	1.22	-1.81	-1.71
0.14183	0.19	-4.96	7.67	3.31
0.28340	0.08	-3.68	1.65	0.43
0.48793	-0.02	1.43	-2.95	-0.58
0.70250	-0.04	4.13	0.85	0.15
0.74950	-0.02	2.21	6.40	1.09
0.89365	0.04	-7.36	3.94	0.61

* denotes deviation in %, per equation 11

Table 6. System: N-Octane/Sulfolane Experimental Vapor-Liquid Equilibrium Data

tC LIQUID	tC VAPOR	P Torr	X1	Y1	γ_1	γ_2	$\log\gamma_1/\log\gamma_2$
98.8	99.4	1.13	0.0	0.0	-	1.0000	-
98.9	99.4	332.3	0.01850	0.99655	53.09	1.0292	1.7125
98.8	99.4	332.3	0.98303	0.99655	1.0024	59.86	-1.7761
98.7	99.3	333.9	0.99548	0.99772	1.0051	64.99	-1.8106
98.9	99.5	336.8	0.99843	0.99897	1.0025	67.64	-1.8291
98.7	99.5	333.0	0.99858	0.99964	0.9954	68.05	-1.8348
98.7	99.4	339.0	0.99858	0.99969	1.0132	65.95	-1.8135
98.8	98.9	336.0	1.00000	1.00000	1.0000	-	-
122.7	122.8	4.0	0.0000	0.0000	-	1.0000	-
122.7	122.8	491.0	0.0197	0.9915	35.26	1.0692	1.5182
122.7	123.0	605.9	0.0252	0.9938	33.93	0.9630	-1.5477
122.6	122.9	691.0	0.9790	0.9952	1.0050	39.859	-1.5983
122.6	123.3	677.0	0.9895	0.9975	0.9906	41.691	-1.6242
122.5	122.9	689.0	0.9953	0.9987	0.9929	48.372	-1.6877
122.3	122.7	682.7	0.9985	0.9996	0.9910	47.039	-1.6764
122.7	122.7	701.0	1.0000	1.0000	1.0000	-	-

Table 7. Coefficients for Correlating Activity Coefficients System: N-Octane/Sulfolane

TEMPERATURE	98.8 C	122.8 C
CORRELATION COEFFICIENTS		
g'12	3.616	
g"12	-0.189	
gamma 12 INF	59.7	39.9
gamma 21 Inf	69.0	47.9
Margules A^M_{12}	1.9023	1.7593
B^M_{12}	-0.1260	-0.1584
A^M_{21}	1.7133	1.5217
B^M_{21}	0.1260	0.1585
Van Laar A_{12}	1.789	1.601
A_{21}	1.839	1.680
NRTL G_{12}	--	--
G_{21}	--	--
α_{12}	--	--

Table 8. Predicted Compared with Experimental VLE Values System: N-Octane/Sulfolane

tC	X1	Y1 EXP.	Y1 CALC.	Y2 CALC.	P (Torr) EXP.	P (Torr) CALC.
98.8	0.01850	0.99655	0.99664	0.00336	332.3	332.35
	0.98303	0.99655	0.99657	0.00343	332.3	331.91
	0.98956	0.99772	0.99778	0.00222	333.9	332.34
	0.99548	0.99897	0.99899	0.00101	336.8	336.01
	0.99843	0.99964	0.99964	0.00036	333.0	334.54
	0.99858	0.99969	0.99968	0.00032	339.0	334.94
122.7	0.0197	0.9915	0.99193	0.00807	491.0	484.2
	0.0252	0.9938	0.99343	0.00655	605.9	597.3
	0.9790	0.9952	0.99511	0.00489	691.0	688.8
	0.9895	0.9975	0.99739	0.00261	677.0	683.8
	0.9953	0.9987	0.99877	0.00123	689.6	694.5
122.3	0.9985	0.9996	0.99960	0.00040	682.0	688.2

Table 9. Deviations Between Predicted and Experimental VLE Values System: N-Octane/Sulfolane

	X1	DY(1)*	DY(2)*	DELP (Torr)	% DELP
t=98.8 C	0.01850	0.009	-2.72	0.05	0.016
	0.98303	0.002	-0.50	-0.39	-0.119
	0.98956	0.006	-2.72	-1.56	-0.467
	0.99548	0.002	-1.86	-0.84	-0.248
	0.99843	0.000	-0.63	1.54	0.463
	0.99858	-0.001	4.50	-4.42	-1.304
t=122.7 C	0.0197	0.04	-5.04	-6.74	-1.37
	0.0252	-0.04	5.57	-8.64	-1.42
	0.9790	-0.01	1.84	-2.17	-0.31
	0.9895	-0.11	4.46	6.83	1.01
	0.9953	0.01	-5.39	4.96	0.72
	0.9985	0.00	-0.36	6.23	0.91

* - denotes deviation in %, per equation 11.

Table 10. System: Ethyl Cyclohexane/Sulfolane Experimental Vapor-Liquid Equilibrium Data

t C LIQUID	t C VAPOR	P Torr	X1	Y1	γ_1	γ_2	$\log \gamma_1/\gamma_2$
104.6	105.1	1.49	0.0	0.0	-	1.0000	-
104.6	105.1	343.9	0.03055	0.9958	37.27	1.0062	1.5061
104.7	105.0	346.0	0.98816	0.9969	1.0019	60.065	-1.7778
104.7	105.3	346.0	0.98894	0.9969	0.9978	63.95	-1.8067
104.7	105.1	348.2	0.98894	0.9971	1.0043	60.14	-1.7773
104.7	104.8	346.0	0.99160	0.9971	1.0010	62.17	-1.7932
104.7	105.4	351.1	0.99180	0.9977	1.0105	65.78	-1.8136
104.5	105.3	347.6	0.99596	0.9988	1.0036	69.73	-1.8419
104.6	104.6	348.6	1.0000	1.0000	-	-	-
132.0	132.1	5.95	0.0000	0.0000	-	1.0000	-
131.9	132.4	755.2	0.0472	0.99252	20.83	1.0044	1.3165
132.0	132.4	756.4	0.98215	0.99485	1.0023	36.82	-1.5651
132.0	132.4	756.4	0.98343	0.99490	1.0010	39.28	-1.5938
132.0	132.5	758.0	0.98579	0.99600	1.0045	36.00	-1.5555
132.0	132.3	760.5	0.99116	0.99716	1.0035	41.22	-1.6147
132.0	132.4	760.0	0.99590	0.99870	1.0023	40.65	-1.6104
132.0	132.3	764.0	1.0000	1.0000	1.0000	-	-

Table 11. Coefficients for Correlating Activity Coefficients System: Ethyl Cyclohexane/Sulfolane

TEMPERATURE	104.6 C	132.0 C
CORRELATION COEFFICIENTS		
g'12	3.4296	3.0675
g"12	-0.7607	-0.5943
gamma 12 INF	38.72	27.21
gamma 21 Inf	70.14	42.93
Margules A^M_{12}	2.0952	1.8309
B^M_{12}	-0.5071	-0.3962
A^M_{21}	1.3345	1.2366
B^M_{21}	0.5071	0.3962
Van Laar A_{12}	1.5880	1.4347
A_{21}	1.8416	1.6328
NRTL G_{12}	---	---
G_{21}	---	---
α_{12}	---	---

Table 12. Predicted Compared with Experimental VLE Values System: Ethyl Cyclohexane/Sulfolane

t C	X1	Y1 EXP.	Y1 CALC.	Y2 CALC.	P(Torr) EXP.	P(Torr) CALC.
104.6	0.03055	0.99580	0.99580	0.00420	343.9	342.50
	0.98816	0.99694	0.99684	0.00316	346.0	345.64
	0.98894	0.99694	0.99702	0.00298	346.0	346.90
	0.98894	0.99714	0.99702	0.00298	348.2	346.90
	0.98916	0.99710	0.99708	0.00292	346.0	345.87
	0.99180	0.99770	0.99773	0.00227	351.1	347.56
	0.99596	0.99880	0.99885	0.00115	347.6	346.38
132.0	0.04720	0.99252	0.99252	0.00748	755.2	756.61
	0.98215	0.99485	0.99483	0.00517	756.4	755.71
	0.98343	0.99490	0.99516	0.00484	756.4	756.31
	0.98579	0.99600	0.99574	0.00426	758.0	757.44
	0.99116	0.99716	0.99725	0.00275	760.5	760.03
	0.99590	0.99870	0.99868	0.00132	760.0	762.37

Table 13. Deviations Between Predicted and Experimental VLE Values System: Ethyl Cyclohexane/Sulfolane

	X1	DY(1)*	DY(2)*	DELP (Torr)	%DELP
t=104.6 C	0.03055	-0.00	0.079	-1.395	0.41
	0.98816	-0.01	3.337	-0.383	-0.11
	0.98894	0.01	-2.574	0.876	0.25
	0.98894	-0.01	4.237	-1.264	-0.36
	0.98914	-0.00	0.715	-0.155	-0.04
	0.99180	0.00	-1.437	-3.542	-1.01
	0.99596	0.00	-3.874	-1.223	-0.35
t=132.0 C	0.04720	0.000	-0.023	1.42	0.19
	0.98215	-0.003	0.592	-0.68	-0.09
	0.98343	0.025	-4.781	-0.08	-0.01
	0.98579	-0.024	6.007	-0.56	-0.07
	0.99116	0.009	-3.213	-0.47	-0.06
	0.99590	-0.002	1.718	2.37	0.32

* - denotes deviation in %, per equation 11.

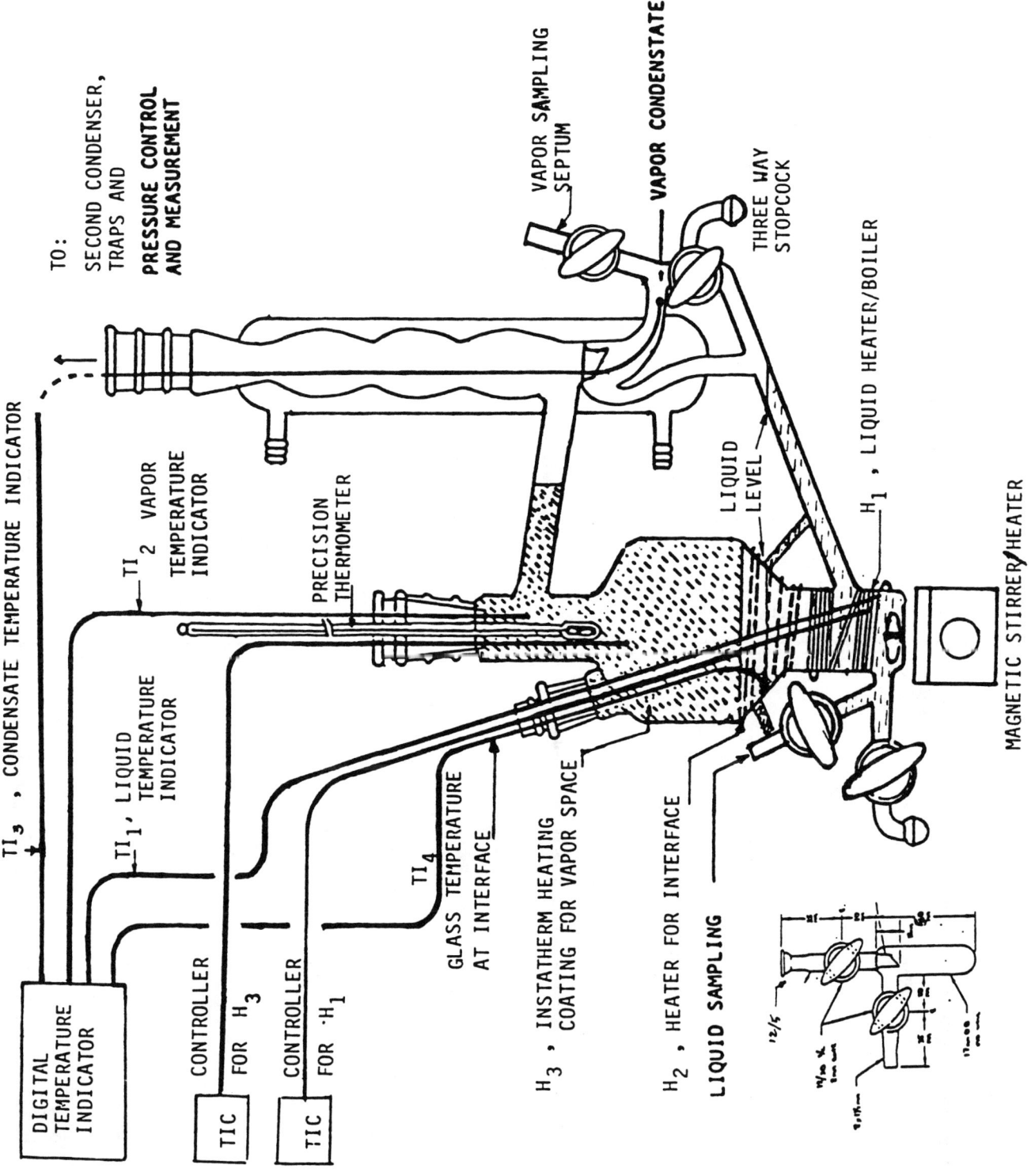

Figure 1. Vapor/Liquid Equilibrium Cell

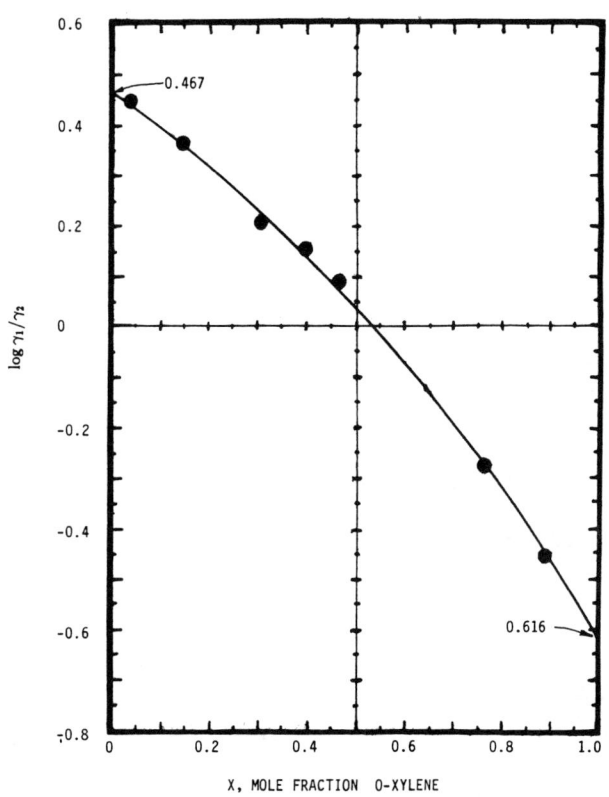

Figure 2. System: O-Xylene/Sulfolane at 110°C
Thermodynamic Consistency Test

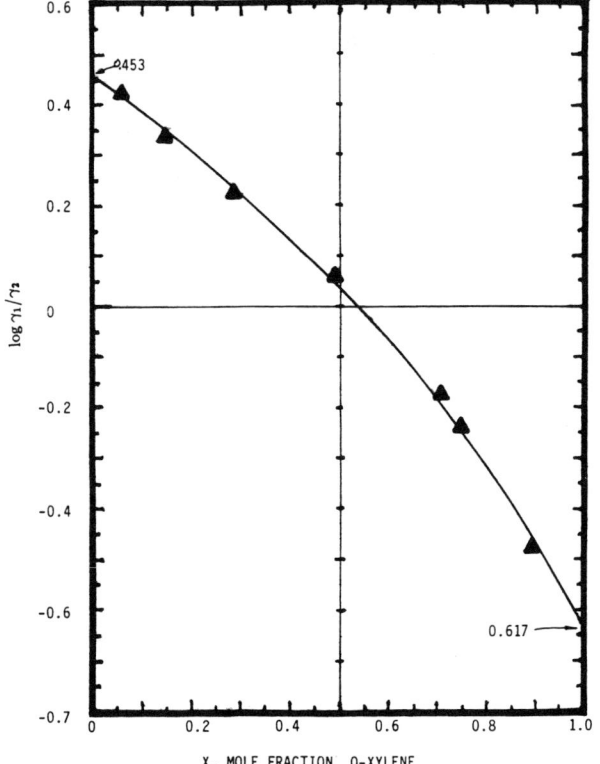

Figure 3. System: O-Xylene/Sulfolane at 142°C
Thermodynamic Consistency Test

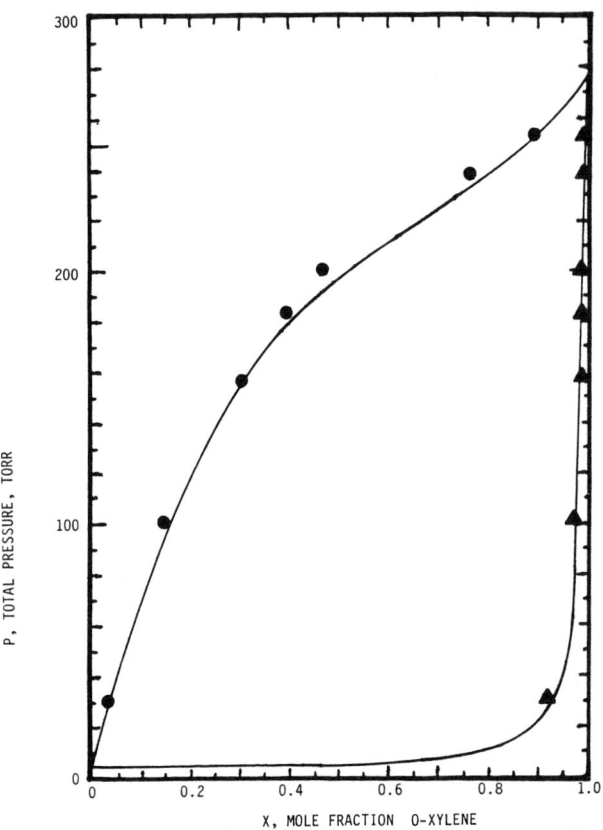

Figure 4. System: Ortho Xylene/Sulfolane at 110°C Total Pressure versus Composition

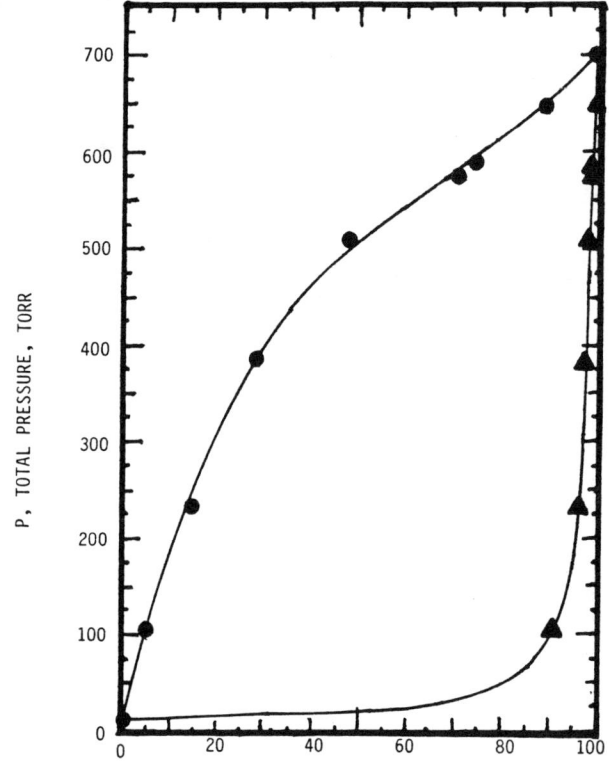

Figure 5. System O-Xylene/Sulfolane at 142°C
Total Pressure versus Composition

VAPOR-LIQUID EQUILIBRIUM MEASUREMENTS ON MIXTURES IMPORTANT TO INDUSTRIAL DESIGN: PHENOL + DIMETHYLETHER, PHENOL + PROPYLAMINE, ETHYLFORMATE + DIMETHYLETHER, ETHYLBENZENE + PROPYLAMINE, METHYLAMINE + DIMETHYL-ETHER, AND ANILINE + DIMETHYLETHER: DIPPR PROJECT 805 NSF (B)/89

Richard L. Rowley and Roger H. Powell ■ Department of Chemical Engineering, Brigham Young University, Provo, UT 84602

Vapor-liquid equilibrium measurements are reported for six binary mixtures of industrial interest on two isotherms. Total pressure measurements were made on mixtures of phenol + dimethylether, phenol + propylamine, ethylformate + dimethylether, ethylbenzene + propylamine, methylamine + dimethylether, and aniline + dimethylether. Results reported include equilibrium phase compositions, fugacity coefficients, activity coefficients, and parameters for the NRTL and UNIQUAC activity coefficient models.

DIPPR project 805/NSF(B)/89 is part of a continued experimental program of vapor-liquid equilibrium (VLE) measurements on binary mixtures important for industrial design and performance evaluation. The systems for this project were chosen by the DIPPR steering committee based on: (1) the need for new and reliable data for certain systems key to processing important industrial chemicals of industry-wide interest and (2) the need for experimental phase behavior on binary mixtures representing mixing behavior specific to interactions of members of different homologous series with different functional groups. In particular, the second objective aims to fill gaps in the current UNIFAC group contribution tables for use in prediction of activity coefficients. In this paper, total pressure, or PTz, vapor-liquid equilibrium measurements for six binary mixtures are reported on two isotherms. Measurements were made on mixtures of phenol + dimethylether, phenol + propylamine, ethylformate + dimethylether, ethylbenzene + propylamine, methylamine + dimethylether, and aniline + dimethylether. Specifically, VLE data on these systems can be used to obtain new UNIFAC parameters for the interactions CH_2O-ACOH, CH_2O-CNH_2, CH_2O-HCOO, CH_2O-$ACNH_2$, $CHNH_2$-ACOH, and $CHNH_2$-$ACCH_2$. Table 1 identifies the isotherms studied for each system.

TABLE 1. Systems and Conditions of Measurements.

Components	Isotherms (°C)	
phenol + dimethylether	46	80
phenol + propylamine	80	160
ethylformate + dimethylether	0	80
ethylbenzene + propylamine	80	160
methylamine + dimethylether	0	80
aniline + dimethylether	40	90

APPARATUS AND MEASUREMENT METHODS

The total pressure (PTz) method is a fast, efficient method to obtain binary vapor-liquid equilibrium data. It eliminates difficulties associated with sampling and analysis common in Pxy and Txy methods. However, accurate pressure measurements at well-controlled temperatures are essential since all equilibrium compositions and related properties (e.g., activity coefficients, fugacity coefficients, etc.) are calculated rather than directly measured. The most common cause of errors in measured pressures is the

TABLE 2. Pure Component Properties.

	phenol	dimethyl-ether	propyl-amine	ethyl-formate	ethyl-benzene	methyl-amine	aniline
Abbrev.	PHE	DME	PAM	EFO	EBE	MAM	ANI
M (g/mol)	94.113	46.069	59.112	74.08	106.168	31.058	93.13
T_c (K)	694.2	400.0	497.0	508.5	617.1	430.0	699.0
P_c (MPa)	6.13	5.37	4.74	4.74	3.61	7.46	5.31
ω	0.440	0.192	0.229	0.285	0.301	0.275	0.384
ρ (g/cc)	1.059	0.667	0.719	0.927	0.867	0.703	1.022
@ T (°C)	40	20	20	16	20	-13.6	20

presence of inerts, so careful de-gassing before each measurement is essential.

For simplicity of discussion, the components will be identified hereafter by the three letter abbreviations shown in Table 2. Table 2 also lists pure-component constants (1, 2) required for the data analysis.

The majority of the measurements were made in a stainless-steel static cell. Pure component vapor pressures were generally measured in the same cell as part of the measurements performed on each system. However, the vapor pressures of PHE and ANI were sufficiently small that they were measured separately using a glass, full-reflux still. An attached manometer referenced to a McLeod gauge was used for the pressure measurements. Measurements at the PHE-rich end of the 80°C PHE + PAM isotherm were made in a glass static cell. This "flask-in-a-bath" apparatus with its attached manometer permitted both the flask and the manometer to be thermostatted at 80°C—a necessary precaution to prevent solidification of PHE in the manometer arm.

The stainless-steel static cell and accompanying apparatus is shown in Figure 1. The cell was cleaned and evacuated before introduction of a weighed charge of the first component. The cell was immersed in a constant temperature bath controlled to ±0.01°C. Thermal equilibration and temperatures within the cell were monitored with a copper-constantan thermocouple probe (referenced to an ice-point bath) which extended into the liquid phase (cf. Figure 1). The thermocouples were made from Omega high-purity, thermocouple wire by welding a measurement and reference junction. Calibration of the thermocouples was checked against a quartz thermometer which has a calibration traceable to IPTS-68.

After thermal equilibration following each component addition, the cell was manually agitated to insure phase equilibrium, the pressure noted, and a degas taken into a previously evacuated and weighed 0.300 L cell. The weight of the degas was recorded and equilibrium was again established by vigorous agitation. The final pressure was recorded after one or more degas samples had been removed to insure removal of all inert gases introduced either in the initial or added charges. A gravimetrically measured amount of the second component was then introduced into the cell, and the above equilibration, degas, and measurement procedures were repeated for the new nominal composition. Sequential additions of the second component were made until the nominal composition was at least 55 mole percent in the

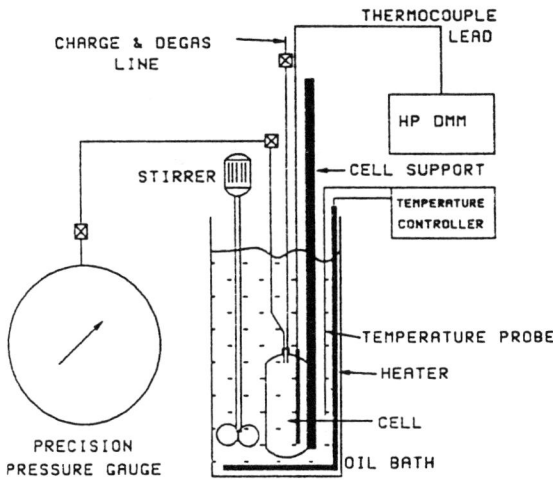

Figure 1. Stainless Steel Static Cell

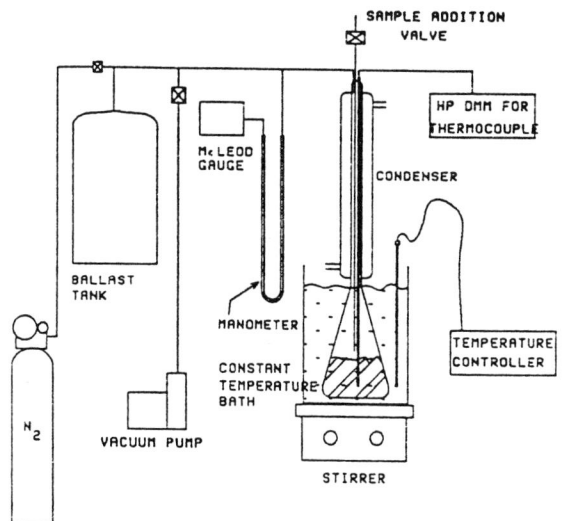

Figure 2. Glass Reflux Cell

second component. The cell was then cleaned, evacuated and gravimetrically charged with the next starting composition. For systems in which the relative volatility of the components were not extremely different, the composition range was covered in two or more sets of experiments starting from each of the pure components. Systems containing DME were only scanned in one compositional direction because of DME's high vapor pressure which made injections of the less volatile component into the cell at high pressures difficult. In all cases, there were compositional regions of overlap between different data sets, and redundant experiments were run to ensure consistency and reliability of the measurements.

All charges, additions, and degas samples were weighed to ±0.01 g. Pressures in the range of 0 to 34.5 kPa (5 psi) were measured using a mercury manometer and are accurate to ±0.014 kPa (0.002 psi). Pressures in the range 34.5 to 41 kPa (5 to 60 psi) were measured with a precision 60 psi Heise gauge calibrated with a dead weight tester, and are believed to be accurate to ±0.07 kPa (0.01 psi). Pressures in the range 0.414 to 4.14 MPa (60 to 600 psi) were measured using a precision 600 psi Heise gauge, also calibrated with a dead-weight tester, and are believed to be accurate to ±0.7 kPa (0.1 psi).

As mentioned, mixtures of PHE + PAM at 80°C containing between 100 % and 70 % PHE were measured in a glass apparatus similar to the stainless steel static cell. This cell had a manometer tube directly attached to the cell itself so that the manometer was co-thermostatted with the measurement cell directly in the constant temperature bath. This permitted measurement of system pressures too low for accurate measurement with guages and eliminated any problem of vapor condensation in the measurement arm of the manometer. Stirring was done with a magnetic stirrer rather than manual agitation, but otherwise the procedure was similar to that employed with the stainless-steel cell.

The glass reflux still shown in Figure 2 was used to measure the pure component vapor pressure of PHE and ANI due to their relatively low vapor pressure. The pure component was charged to the clean, evacuated still and thermostatted in a constant temperature bath at a set point about 10°C above the desired temperature in order to maintain an appreciable boil-up rate. The pressure was adjusted until steady reflux occurred at the desired temperature and the pressure was constant. This pressure was recorded as the pure component vapor pressure at that temperature. For PHE and ANI, vapor pressures were measured over a range of temperatures and a fit was made of the $\ln(P°)$ vs. T^{-1} data. Antoine constants were regressed from this plot and used in the case of PHE to extrapolate the vapor pressure to the lower isotherm where accurate measurements of such a small vapor pressure would have been impossible. Pressures were measured with a mercury manometer referenced to near vacuum conditions, as monitored with a McLeod gauge accurate to 1 μm Hg.

A measuring cathetometer was used to read Hg levels in the manometer to ±0.1 mm.

CHEMICALS

All chemicals were used directly as received from the manufacturer without further purification. The PHE, and ANI were obtained from Mallinckrodt Chemical Co. with stated minimum purities of 99.45% and 99.6%, respectively. PAM and EBE were supplied by Aldrich Chemical with purities of 98% and 99+%, respectively. DME and MAM were supplied by Matheson Gas at 99% purity. EFO was obtained from MCB Manufacturing Chemists with a stated purity of 98%.

DATA REDUCTION

Measured pressures for each nominal composition on an issotherm were reduced to equilibrium phase compositions using the above modification of the Redlich-Kwong equation of state to calculate the vapor fugacity coefficients and the NRTL and UNIQUAC models for liquid-phase activity coefficients. Phase equilibrium requires

$$f_i^V = f_i^L \qquad (1)$$

where f_i^V and f_i^L are the fugacities of component i in the vapor and liquid phases, respectively. The vapor fugacity is given by

$$f_i^V = y_i P \phi_i \qquad (2)$$

where P is pressure, y_i, is vapor mole fraction of component i, and ϕ_i is the fugacity coefficient of component i. The liquid fugacity is given by

$$f_i^L = f_i^o x_i \gamma_i \qquad (3)$$

where f_i^o is the standard state fugacity, x_i is the liquid mole fraction, and γ_i is the activity coefficient of component i.

The standard state fugacity is defined as the fugacity of pure component i at the system temperature and pressure. It is calculated from the saturated fugacity of component i at the system temperature, $f_{i,sat}^o$, corrected to the system pressure with the known pressure dependence of the fugacity coefficient in a so-called Poynting correction factor. Since at saturation the pure component fugacity coefficient can be written as

$$f_{i,sat}^o = \phi_i^o P_i^o \qquad (4)$$

where ϕ_i^o is the fugacity coefficient at the system temperature and at the corresponding saturation or vapor pressure of component i, P_i^o. Therefore,

$$f_i^o = f_{i,sat}^o \exp\left(\int_{P_i^o}^{P} \frac{\overline{V}}{RT} dP\right) \qquad (5)$$

where $\overline{V}$ is the partial molal volume of component i.

Combining Equations (2) through (5) yields the final form of the VLE analysis equation,

$$y_i P \phi_i = x_i \gamma_i \phi_i^o P_i^o \exp\left(\int_{P_i^o}^{P} \frac{\overline{V}}{RT} dP\right) \qquad (6)$$

Calculation of equilibrium vapor compositions from *PTz* data requires an iterative procedure based on Equation (6). In essence, the procedure is a regression of liquid and vapor compositions by comparison of measured and calculated pressures for all measurements on the isotherm simultaneously. The specifics of this process are shown in the simplified flow diagram of Figure 3. The iterations are begun by assuming the liquid compositions to be the same as the charge compositions and the fugacity coefficients to be unity. This allows activity coefficients to be calculated at the prescribed temperature. Equation (6) can then be used to solve for the product y_iP for each component. The total calculated pressure is then given by $P_{tot} = \Sigma(y_iP)$. With updated vapor compositions, the next iteration corrects the liquid compositions for the vaporized material and the degas using the known volume of the cell and material balances. This iteration loop is terminated upon satisfactory agreement between calculated and measured pressures. The above procedure is repeated at each measurement point.

RESULTS

Pure component molecular weights, critical constants, acentric factors, liquid densities, and vapor pressures are required for the analysis method shown above. Values for these constants were available in the literature [1, 2]; however, vapor pressures were measured as part of this study. Because the vapor pressures of PHE at 46°C and ANI to 40°C were too

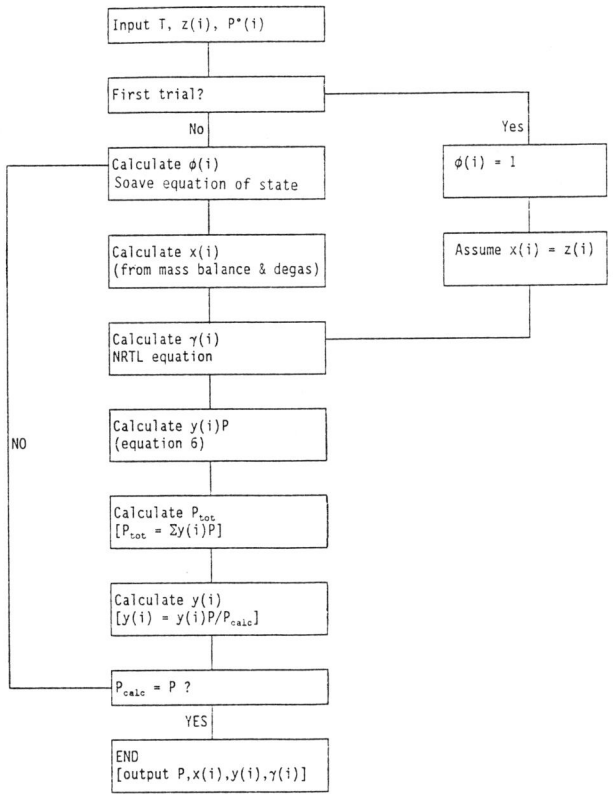

Figure 3. Flow Chart for VLE Data Reduction

TABLE 3. Pure Component Vapor Pressure for PHE.

T (K)	P° (kPa)
353.23	1.839
361.93	3.012
372.86	5.224
382.56	8.083
392.77	12.41
402.90	18.64
412.54	26.60
422.65	38.01
432.53	52.22

$A = 14.5125$
$B = 3465.29$
$C = -104.145$

TABLE 4. Pure Component Vapor Pressure for ANI.

T (K)	P° (kPa)
349.53	1.719
365.30	3.985
375.36	6.285
385.09	9.518
394.26	13.70
405.04	20.30
415.59	29.27
425.22	40.86
433.58	52.75
443.64	70.12

$A = 14.2466$
$B = 3502.62$
$C = -93.096$

low for accurate measurement, vapor pressures were measured from 80.08°C to 159.31°C for pure PHE and from 75.51°C to 170.49°C for pure ANI. Antoine constants were regressed from these data and the Antoine equation

$$\ln P° = A - \frac{B}{T + C} \qquad (7)$$

($P°$ in kPa and T in K)

was used to extrapolate the pure component vapor pressures at 40°C. Tables 3 and 4 show the results of the PHE and ANI vapor pressure studies and the resultant Antoine constants. The agreement between the measured data and the fitted Antoine equation is shown in Figures 4 and 5.

Tables 5 - 16 contain the experimental and calculated results for the twelve systems measured in this study. The *PTz* analysis requires use of an activity coefficient model to describe the liquid mixture nonidealities over the entire composition range. Temperature, pressure and charge composition are the raw experimental data from which the analysis program calculates the liquid and vapor compositions. In general, the analysis was performed twice, once with the NRTL (3) model and once with the UNIQUAC (4) model for the activity coefficients. However, mixtures containing PHE were sufficiently nonideal that a four parameter Redlich-Kister (RK) model was required to adequately represent the data. The Soave (5) modification of the Redlich-Kwong equation of state was used in both cases to calculate

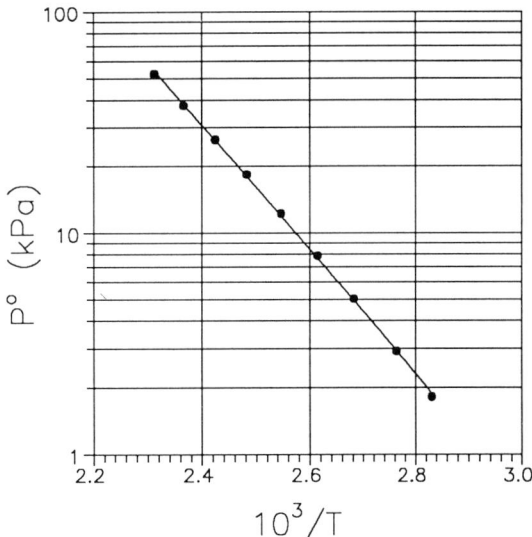

Figure 4. Measured PHE Vapor Pressures (•) and Correlation of Data with Antoine Equation.

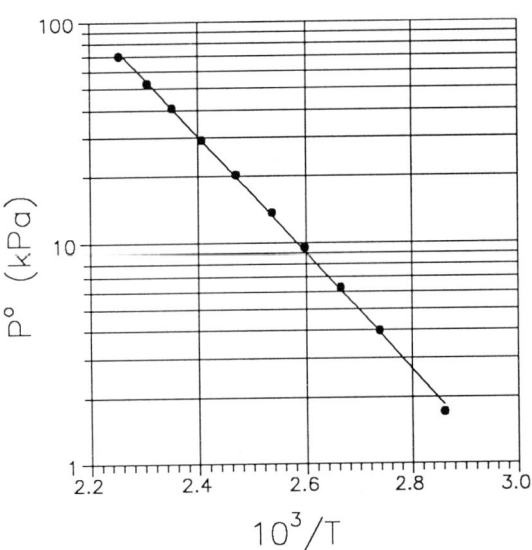

Figure 5. Measured ANI Vapor Pressures (•) and Correlation of Data with Antoine Equation.

vapor phase nonidealities. In addition to the measured temperature and pressure conditions, also shown in these tables of results are the equilibrium compositions, activity coefficients, and fugacity coefficients obtained from the analysis using the two different models. A weighted nonlinear least squares program was used in conjunction with the analysis program to obtain a best fit of the experimental data by adjusting the parameters in the activity coefficient model. Final values for the parameters are shown in Table 17.

Figures 6-17 depict the quality of the data and the efficacy of the activity coefficient model. Figure 6-9 show the results obtained with the RK correlation and Figures 10-17 are for the NRTL model. Measured pressures are shown as solid circles, calculated values as continuous lines. Also shown in these figures are plots of the equilibrium x-y values and the percent deviation of measured pressures from Raoult's law. These latter plots emphasize the system nonidealities by subtracting the ideal contribution as background from the measured pressures. Thus,

$$\% \text{ Dev.} = 100\% \times (P_{meas} - P_{id})/P_{id} \qquad (8)$$

where

$$P_{id} = P_1^\circ + (P_2^\circ - P_1^\circ)x_2 \qquad (9)$$

Any inconsistencies among the several experimental sets that were run for each system and between individually measured points show up in these plots as scattered deviations from a curve. However, because P_{id} varies considerably, an absolute comparison cannot be made from these plots, and only relative trends are significant. For example, at compositions where the system contains mostly a relatively nonvolatile component, P_{id} may be quite small causing the % deviation calculated from Equation 7 to appear quite large. Therefore, experimental noise in a measurement in this region will cause a significantly larger scatter in these deviation plots than the same amount of noise at a composition where the volatile component predominates P_{id}.

The pure component vapor pressures for many of these systems were different enough that the vapor phase was composed of almost exclusively the more volatile component over the entire composition range as can be seen from the x-y plots. Only in the case of the MAM + DME system at 80°C did we measure an azeotrope. A low boiling azeotrope occurs at this temperature in this system at approximately 90% DME (see Figure 14). At low PAM concentrations in PHE + PAM mixtures, the vapor pressure is extremely flat with respect to composition (see

Figures 8 and 9). The pressure does rise slightly, but continuously, with increasing PAM concentrations. The two parameter UNIQUAC and NRTL models cannot adequately represent the data over the entire composition range. If used, they predict an azeotrope in the low pressure range in order to adequately represent the high-pressure end. The increased flexibility of the four parameter RK model eliminates this problem. Although the UNIQUAC and NRTL fits may be adequate for the PHE + DME system, the RK results agree somewhat better at the low pressure compositions as shown in Tables 5 and 6.

SUMMARY

PTz measurements have been made on the PHE + DME, PHE + PAM, EFO + DME, EBE + PAM, MAM + DME, and ANI + DME systems each at two different temperatures. The results include equilibrium vapor and liquid mole fractions, calculated activity and fugacity coefficients, and regressed parameters for NRTL, UNIQUAC or RK activity coefficient models. Agreement of results between replicate and continuation experiments in regions of compositional overlap indicate good internal consistency of the data.

LITERATURE CITED

1. Daubert, T.E., and R.P. Danner, *DIPPR Project 801 Data Compilation, Tables of Physical and Thermodynamic Properties of Pure Compounds*, Extant 1990.

2. Reid, R.C., J.M. Prausnitz and B.E. Poling, *The Properties of Liquids and Gases*, 4th ed., McGraw-Hill, New York, 1987.

3. Renon, H., and J.M. Prausnitz, *AIChE J.* **14**, 135 (1968).

4. Abrams, D.S., and J.M. Prausnitz, *AIChE J.* **21**, 116 (1975).

5. Soave, G., *Chem. Eng. Sci.* **27**, 1197 (1972).

ACKNOWLEDGEMENT

Support of this work under DIPPR project 805 NSF(B) is gratefully acknowledged.

TABLE 5. VLE Results for PHE + DME at 80°C.

Mole Percent PHE		P (KPa)		Activity Coeff		Fugacity Coeff	
Liquid	Vapor	Meas	Calc	PHE	DME	PHE	DME
(A) NRTL							
100.000	100.000	1.8	1.8	1.000	0.176	0.999	1.000
86.203	1.935	84.5	80.5	0.954	0.343	0.974	0.992
80.876	1.064	125.6	134.8	0.919	0.415	0.957	0.987
75.255	0.629	198.5	207.7	0.876	0.492	0.934	0.981
70.950	0.442	264.8	274.3	0.841	0.550	0.914	0.974
64.077	0.267	376.5	399.2	0.783	0.638	0.876	0.963
59.319	0.195	483.9	498.0	0.743	0.695	0.847	0.954
50.937	0.118	686.0	693.5	0.674	0.784	0.791	0.936
46.911	0.094	801.8	796.0	0.642	0.821	0.762	0.926
42.258	0.072	930.5	920.2	0.607	0.859	0.728	0.915
41.770	0.071	925.1	933.5	0.603	0.862	0.725	0.913
37.752	0.057	1035.3	1045.5	0.575	0.891	0.695	0.903
36.065	0.052	1122.3	1093.6	0.563	0.901	0.682	0.899
30.562	0.038	1308.6	1254.1	0.527	0.932	0.640	0.884
29.676	0.036	1279.8	1280.5	0.521	0.936	0.634	0.882
21.206	0.022	1559.9	1537.5	0.470	0.970	0.570	0.858
17.822	0.017	1674.3	1642.4	0.452	0.979	0.545	0.848
11.140	0.010	1891.4	1852.7	0.418	0.992	0.495	0.829
4.838	0.004	2094.8	2054.1	0.388	0.999	0.449	0.810
0.000	0.000	2210.6	2210.6	0.368	1.000	0.414	0.796
(B) RK							
100.000	100.000	1.8	1.8	1.000	0.262	0.999	1.000
86.203	1.928	84.5	82.9	0.977	0.344	0.972	0.992
80.876	1.133	125.6	131.2	0.951	0.394	0.956	0.988
75.255	0.691	198.5	196.9	0.913	0.456	0.935	0.982
70.950	0.487	264.8	258.8	0.878	0.508	0.916	0.976
64.077	0.291	376.5	380.1	0.812	0.596	0.878	0.964
59.319	0.208	483.9	480.2	0.764	0.658	0.847	0.955
50.937	0.121	686.0	685.5	0.679	0.762	0.786	0.936
46.911	0.094	801.8	794.9	0.640	0.806	0.754	0.926
42.258	0.072	930.5	927.6	0.597	0.852	0.717	0.914
41.770	0.070	925.1	941.8	0.593	0.856	0.713	0.912
37.752	0.056	1035.3	1060.6	0.560	0.889	0.680	0.901
36.065	0.051	1122.3	1111.1	0.547	0.902	0.667	0.897
30.562	0.037	1308.6	1277.3	0.508	0.936	0.623	0.881
29.676	0.036	1279.8	1304.1	0.502	0.941	0.616	0.879
21.206	0.022	1559.9	1560.8	0.454	0.974	0.551	0.855
17.822	0.017	1674.3	1663.2	0.438	0.982	0.526	0.846
11.140	0.010	1891.4	1865.9	0.410	0.994	0.477	0.827
4.838	0.004	2094.8	2059.6	0.386	0.999	0.432	0.809
0.000	0.000	2210.6	2210.6	0.367	1.000	0.398	0.795

TABLE 6. VLE Results for PHE + DME at 46°C.

Mole Percent PHE		P (KPa)		Activity Coeff		Fugacity Coeff	
Liquid	Vapor	Meas	Calc	PHE	DME	PHE	DME
(A) NRTL							
100.000	100.000	0.2	0.2	1.000	0.102	1.000	1.000
81.087	0.275	47.0	54.0	0.894	0.326	0.977	0.993
75.949	0.157	73.4	84.8	0.844	0.401	0.964	0.989
65.991	0.064	143.2	166.0	0.744	0.548	0.931	0.979
58.007	0.035	229.7	249.3	0.665	0.659	0.897	0.969
50.950	0.021	325.6	333.9	0.600	0.745	0.864	0.958
46.628	0.016	387.1	389.4	0.563	0.792	0.842	0.951
40.009	0.011	488.1	478.5	0.511	0.853	0.808	0.940
35.259	0.008	565.4	544.4	0.476	0.890	0.783	0.932
30.540	0.006	635.3	610.7	0.445	0.921	0.759	0.923
30.060	0.006	643.6	617.4	0.441	0.924	0.756	0.922
25.486	0.004	710.0	681.8	0.413	0.947	0.733	0.914
21.363	0.003	764.1	739.5	0.390	0.964	0.712	0.907
20.164	0.003	786.0	756.2	0.384	0.968	0.706	0.905
16.472	0.002	833.2	807.1	0.364	0.980	0.688	0.899
9.997	0.001	920.2	894.5	0.334	0.993	0.658	0.888
4.803	0.001	985.3	962.3	0.312	0.998	0.634	0.879
0.000	0.000	1023.0	1023.0	0.293	1.000	0.614	0.872
(B) RK							
100.000	100.000	0.2	0.2	1.000	0.102	1.000	1.000
81.087	0.327	47.0	47.2	0.932	0.326	0.979	0.994
75.949	0.192	73.4	72.5	0.885	0.401	0.969	0.991
65.991	0.075	143.2	145.3	0.770	0.548	0.938	0.982
58.007	0.038	229.7	229.6	0.667	0.659	0.902	0.971
50.950	0.021	325.6	322.5	0.579	0.745	0.864	0.959
46.628	0.015	387.1	386.0	0.529	0.792	0.839	0.951
40.009	0.010	488.1	489.2	0.462	0.853	0.798	0.938
35.259	0.007	565.4	564.5	0.421	0.890	0.769	0.929
30.540	0.005	635.3	637.7	0.388	0.921	0.742	0.920
30.060	0.005	643.6	645.0	0.385	0.924	0.739	0.919
25.486	0.004	710.0	712.3	0.360	0.947	0.714	0.910
21.363	0.003	764.1	769.1	0.344	0.964	0.693	0.903
20.164	0.003	786.0	784.9	0.341	0.968	0.688	0.901
16.472	0.002	833.2	831.6	0.333	0.980	0.671	0.895
9.997	0.001	920.2	907.8	0.332	0.993	0.644	0.886
4.803	0.001	985.3	966.5	0.344	0.998	0.623	0.878
0.000	0.000	1023.0	1023.0	0.369	1.000	0.604	0.871

TABLE 7. VLE Results for PHE + PAM at 160°C.

Mole Percent PHE		P (KPa)		Activity Coeff		Fugacity Coeff	
Liquid	Vapor	Meas	Calc	PHE	PAM	PHE	PAM
(A) RK							
100.000	100.000	52.1	52.1	1.000	0.120	0.987	0.995
82.827	55.238	75.7	75.8	0.963	0.142	0.981	0.992
80.397	48.194	83.3	82.6	0.942	0.157	0.980	0.991
76.864	38.166	96.5	95.8	0.902	0.184	0.977	0.989
75.290	33.949	99.8	103.3	0.881	0.198	0.975	0.989
70.199	22.134	140.9	135.8	0.803	0.253	0.967	0.985
68.546	19.006	147.0	149.6	0.775	0.275	0.964	0.983
63.973	12.181	197.0	197.6	0.694	0.341	0.953	0.978
59.468	7.714	271.0	260.6	0.613	0.417	0.938	0.971
58.677	7.115	278.5	273.4	0.599	0.431	0.935	0.970
58.662	7.104	264.8	273.7	0.599	0.431	0.935	0.970
55.296	5.040	327.8	333.9	0.541	0.494	0.921	0.963
51.850	3.566	401.8	405.1	0.485	0.559	0.905	0.955
50.201	3.031	442.2	442.4	0.460	0.591	0.896	0.951
46.602	2.144	519.6	530.2	0.410	0.659	0.876	0.941
45.490	1.932	553.2	558.8	0.396	0.679	0.869	0.938
41.092	1.297	718.2	677.9	0.345	0.755	0.842	0.925
40.522	1.233	677.8	693.9	0.339	0.764	0.839	0.923
35.615	0.813	876.4	833.9	0.293	0.836	0.807	0.908
35.239	0.788	828.5	844.7	0.290	0.841	0.805	0.906
30.493	0.540	968.8	980.9	0.255	0.895	0.775	0.891
28.482	0.462	1014.8	1037.7	0.243	0.913	0.762	0.885
25.073	0.356	1134.6	1132.0	0.225	0.939	0.741	0.875
21.391	0.268	1212.0	1230.4	0.209	0.960	0.720	0.864
19.894	0.238	1277.7	1269.5	0.203	0.967	0.711	0.860
15.572	0.166	1387.4	1379.2	0.190	0.982	0.687	0.848
10.224	0.097	1527.7	1510.7	0.176	0.993	0.659	0.833
0.000	0.000	1758.0	1758.0	0.154	1.000	0.604	0.806

TABLE 8. VLE Results for PHE + PAM at 80°C.

Mole Percent PHE		P (KPa)		Activity Coeff		Fugacity Coeff	
Liquid	Vapor	Meas	Calc	PHE	PAM	PHE	PAM
(A) RK							
100.000	100.000	1.8	1.8	1.000	0.016	0.999	1.000
87.939	81.179	2.4	2.0	1.002	0.011	0.999	1.000
84.340	71.570	2.5	2.1	0.969	0.014	0.999	1.000
79.603	53.111	2.9	2.5	0.889	0.020	0.999	1.000
74.634	30.841	4.2	3.4	0.766	0.033	0.999	0.999
64.278	5.144	13.9	10.8	0.466	0.102	0.996	0.998
58.370	1.600	20.8	21.8	0.322	0.184	0.991	0.996
54.787	0.798	33.0	32.4	0.253	0.253	0.987	0.993
50.782	0.379	46.7	48.1	0.191	0.346	0.980	0.990
45.073	0.142	79.0	77.7	0.129	0.498	0.968	0.984
40.068	0.066	110.6	108.7	0.093	0.635	0.955	0.978
36.366	0.040	130.5	133.0	0.074	0.727	0.945	0.973
35.090	0.034	156.0	141.4	0.069	0.757	0.942	0.971
30.074	0.019	167.8	173.0	0.054	0.854	0.929	0.965
25.504	0.012	192.1	198.9	0.045	0.915	0.918	0.960
25.066	0.011	197.2	201.2	0.044	0.920	0.917	0.959
19.986	0.007	224.6	225.1	0.038	0.961	0.907	0.954
15.203	0.004	244.1	245.1	0.035	0.981	0.899	0.950
10.116	0.003	265.4	263.8	0.032	0.981	0.892	0.947
0.000	0.000	298.5	298.5	0.027	1.000	0.878	0.940

TABLE 9. VLE Results for EFO + DME at 80°C.

Mole Percent EFO		P (KPa)		Activity Coeff		Fugacity Coeff	
Liquid	Vapor	Meas	Calc	EFO	DME	EFO	DME
(A) NRTL							
100.000	100.000	229.5	229.5	1.000	0.777	0.949	0.984
93.477	70.801	320.1	308.8	0.998	0.814	0.932	0.975
87.860	54.217	377.2	384.7	0.995	0.845	0.917	0.966
83.348	44.386	456.6	450.9	0.990	0.869	0.904	0.959
79.497	37.705	510.7	511.1	0.985	0.889	0.893	0.953
73.237	29.255	648.3	616.1	0.975	0.919	0.873	0.943
69.361	25.139	688.8	685.5	0.967	0.937	0.860	0.936
58.873	16.910	898.3	888.6	0.945	0.977	0.823	0.916
54.134	14.191	1017.5	986.9	0.934	0.992	0.805	0.907
50.013	12.191	1109.3	1075.0	0.925	1.003	0.789	0.898
44.074	9.784	1223.7	1205.3	0.913	1.014	0.766	0.886
40.518	8.561	1307.2	1284.8	0.907	1.019	0.752	0.878
33.348	6.484	1463.3	1446.5	0.899	1.024	0.723	0.863
29.224	5.476	1546.2	1539.7	0.899	1.025	0.707	0.854
24.108	4.369	1663.4	1654.9	0.905	1.022	0.687	0.843
18.849	3.355	1780.4	1772.1	0.920	1.018	0.666	0.832
17.951	3.192	1781.1	1792.5	0.924	1.017	0.663	0.830
12.279	2.198	1931.8	1918.5	0.959	1.010	0.641	0.818
10.877	1.959	1943.5	1950.2	0.972	1.008	0.636	0.815
4.972	0.939	2104.4	2086.4	1.045	1.002	0.612	0.802
0.000	0.000	2210.6	2210.6	1.145	1.000	0.591	0.790
(B) UNIQUAC							
100.000	100.000	229.5	229.5	1.000	0.824	0.949	0.984
93.519	70.235	320.1	311.9	0.999	0.843	0.932	0.975
87.899	53.946	377.2	388.1	0.997	0.860	0.916	0.966
83.367	44.408	456.6	453.1	0.994	0.874	0.904	0.959
79.486	37.932	510.7	511.4	0.991	0.886	0.893	0.953
73.179	29.716	648.3	611.8	0.985	0.905	0.874	0.943
69.270	25.670	688.8	677.6	0.979	0.918	0.862	0.937
58.732	17.438	898.3	870.0	0.961	0.950	0.826	0.918
53.989	14.647	1017.5	964.0	0.950	0.964	0.809	0.909
49.875	12.564	1109.3	1049.4	0.939	0.975	0.794	0.901
43.962	10.018	1223.7	1178.4	0.923	0.991	0.771	0.888
40.426	8.709	1307.2	1258.8	0.912	0.999	0.757	0.881
33.302	6.466	1463.3	1427.2	0.891	1.013	0.727	0.865
29.199	5.378	1546.2	1527.2	0.880	1.019	0.709	0.855
24.100	4.200	1663.4	1653.0	0.870	1.023	0.687	0.843
18.845	3.156	1780.4	1781.4	0.868	1.024	0.665	0.831
17.969	2.998	1781.1	1802.9	0.869	1.023	0.661	0.829
12.286	2.045	1931.8	1935.7	0.896	1.018	0.638	0.816
10.903	1.831	1943.5	1967.1	0.909	1.016	0.633	0.813
4.975	0.927	2104.4	2096.7	1.033	1.006	0.610	0.801
0.000	0.000	2210.6	2210.6	1.337	1.000	0.591	0.790

TABLE 10. VLE Results for EFO + DME at 0°C.

Mole Percent EFO		P (KPa)		Activity Coeff		Fugacity Coeff	
Liquid	Vapor	Meas	Calc	EFO	DME	EFO	DME
(A) NRTL							
100.000	100.000	13.8	13.8	1.000	0.935	0.994	0.998
92.791	42.267	29.1	30.5	0.999	0.952	0.987	0.994
89.401	32.277	45.0	38.6	0.999	0.959	0.984	0.992
81.760	19.991	56.9	57.3	0.996	0.973	0.976	0.989
77.649	16.133	68.1	67.6	0.994	0.980	0.972	0.986
71.058	11.868	84.9	84.4	0.991	0.990	0.965	0.983
67.044	9.992	96.3	94.9	0.989	0.995	0.961	0.981
60.599	7.696	112.3	111.7	0.985	1.001	0.955	0.977
56.483	6.555	122.7	122.6	0.983	1.004	0.950	0.975
51.822	5.481	134.6	135.0	0.981	1.007	0.946	0.973
47.843	4.707	145.1	145.5	0.979	1.009	0.941	0.971
42.203	3.785	162.8	160.5	0.978	1.010	0.936	0.968
38.868	3.316	170.3	169.3	0.977	1.010	0.932	0.966
32.199	2.511	189.5	186.9	0.978	1.010	0.925	0.962
28.384	2.115	198.4	196.9	0.980	1.009	0.921	0.960
22.896	1.609	210.0	211.3	0.985	1.007	0.916	0.957
18.718	1.265	223.7	222.2	0.992	1.005	0.912	0.955
15.548	1.023	230.6	230.4	0.998	1.004	0.908	0.954
10.304	0.653	245.0	244.2	1.011	1.002	0.903	0.951
4.989	0.306	264.8	258.2	1.030	1.001	0.898	0.948
0.000	0.000	268.0	268.0	1.055	1.000	0.893	0.945
(B) UNIQUAC							
100.000	100.000	13.8	13.8	1.000	0.952	0.994	0.998
92.794	42.018	29.1	30.7	1.000	0.962	0.987	0.994
89.402	32.113	45.0	38.8	0.999	0.967	0.983	0.992
81.762	19.960	56.9	57.5	0.997	0.977	0.976	0.989
77.650	16.137	68.1	67.7	0.996	0.982	0.972	0.986
71.058	11.899	84.9	84.4	0.994	0.989	0.965	0.983
67.043	10.028	96.3	94.8	0.992	0.993	0.961	0.981
60.598	7.728	112.3	111.6	0.988	1.000	0.955	0.978
56.482	6.579	122.7	122.5	0.986	1.003	0.950	0.975
51.822	5.496	134.6	134.9	0.983	1.006	0.946	0.973
47.843	4.714	145.1	145.5	0.981	1.009	0.941	0.971
42.204	3.781	162.8	160.7	0.978	1.011	0.935	0.968
38.869	3.307	170.3	169.6	0.976	1.012	0.932	0.966
32.200	2.495	189.5	187.5	0.975	1.013	0.925	0.962
28.385	2.098	198.4	197.7	0.976	1.013	0.921	0.960
23.640	1.658	210.7	210.2	0.979	1.011	0.916	0.958
18.718	1.254	223.7	223.1	0.987	1.009	0.911	0.955
15.549	1.017	230.6	231.3	0.995	1.008	0.908	0.953
10.304	0.655	245.0	244.8	1.017	1.004	0.903	0.951
4.989	0.314	264.8	258.4	1.056	1.001	0.898	0.948
0.000	0.000	268.0	268.0	1.118	1.000	0.893	0.945

TABLE 11. VLE Results for EBE + PAM at 160°C.

Mole Percent EBE		P (KPa)		Activity Coeff		Fugacity Coeff	
Liquid	Vapor	Meas	Calc	EBE	PAM	EBE	PAM
(A) NRTL							
100.000	100.000	189.9	190.0	1.000	1.052	0.946	0.987
90.734	56.832	314.8	315.9	1.000	1.052	0.913	0.970
80.531	36.301	473.2	459.0	1.000	1.049	0.878	0.952
68.737	23.854	648.2	629.7	1.002	1.044	0.839	0.932
58.876	17.411	788.3	776.5	1.006	1.038	0.806	0.915
49.166	12.882	926.2	924.4	1.012	1.030	0.773	0.899
44.611	11.156	959.3	995.0	1.016	1.027	0.758	0.891
39.394	9.404	1092.4	1076.8	1.022	1.022	0.740	0.882
31.832	7.191	1179.1	1197.8	1.033	1.016	0.714	0.868
21.257	4.546	1372.3	1372.5	1.056	1.008	0.678	0.849
10.344	2.146	1564.8	1562.9	1.091	1.002	0.639	0.828
0.000	0.000	1758.0	1757.9	1.139	1.000	0.599	0.806
(B) UNIQUAC							
100.000	100.000	189.9	190.0	1.000	1.064	0.946	0.987
90.740	56.734	314.8	316.6	1.000	1.057	0.913	0.970
80.532	36.334	473.2	459.2	1.002	1.049	0.878	0.952
68.732	23.961	648.2	628.1	1.005	1.040	0.839	0.932
58.867	17.525	788.3	773.0	1.009	1.032	0.806	0.916
49.159	12.974	926.2	919.5	1.015	1.024	0.774	0.899
44.567	11.218	959.3	990.2	1.019	1.021	0.759	0.891
39.389	9.458	1092.4	1071.2	1.025	1.017	0.741	0.882
31.810	7.208	1179.1	1192.6	1.034	1.012	0.716	0.869
21.252	4.532	1372.3	1368.4	1.051	1.006	0.679	0.849
10.344	2.119	1564.8	1560.9	1.077	1.002	0.639	0.828
0.000	0.000	1758.0	1757.9	1.110	1.000	0.599	0.806

TABLE 12. VLE Results for EBE + PAM at 80°C.

Mole Percent EBE		P (KPa)		Activity Coeff		Fugacity Coeff	
Liquid	Vapor	Meas	Calc	EBE	PAM	EBE	PAM
(A) NRTL							
100.000	100.000	22.7	22.7	1.000	0.853	0.989	0.997
90.931	48.204	42.9	43.1	0.997	0.900	0.979	0.992
83.429	30.989	62.2	61.8	0.992	0.935	0.971	0.988
78.093	23.639	75.6	75.9	0.986	0.958	0.965	0.985
72.569	18.295	91.6	91.1	0.979	0.980	0.958	0.982
66.653	14.165	107.5	108.0	0.970	1.000	0.951	0.978
62.913	12.136	120.2	119.0	0.965	1.011	0.947	0.976
55.900	9.173	140.1	139.8	0.954	1.027	0.938	0.972
50.395	7.406	154.6	156.1	0.947	1.036	0.931	0.968
45.525	6.138	170.0	170.5	0.942	1.041	0.925	0.966
41.237	5.200	183.6	183.0	0.940	1.042	0.920	0.963
36.173	4.261	197.3	197.5	0.941	1.042	0.914	0.960
31.232	3.485	210.5	211.1	0.946	1.039	0.908	0.957
28.993	3.170	215.0	217.2	0.951	1.037	0.906	0.956
26.481	2.839	221.5	223.8	0.958	1.034	0.903	0.955
24.215	2.558	228.7	229.7	0.966	1.031	0.901	0.954
21.658	2.260	238.1	236.3	0.978	1.027	0.898	0.952
17.166	1.770	252.2	247.5	1.007	1.020	0.894	0.950
16.804	1.732	245.3	248.4	1.010	1.020	0.893	0.950
12.369	1.281	268.2	259.3	1.053	1.012	0.889	0.948
11.844	1.229	257.2	260.6	1.059	1.012	0.888	0.947
7.081	0.754	269.7	272.3	1.130	1.005	0.883	0.945
0.000	0.000	298.5	298.5	1.299	1.000	0.876	0.941

TABLE 13. VLE Results for MAM + DME at 80°C

Mole Percent MAM		P (KPa)		Activity Coeff		Fugacity Coeff	
Liquid	Vapor	Meas	Calc	MAM	DME	MAM	DME
(A) NRTL							
100.000	100.000	1655.8	1655.8	1.000	1.321	0.859	0.848
95.808	93.477	1701.0	1708.5	1.000	1.292	0.855	0.843
90.700	86.259	1759.3	1768.5	1.002	1.259	0.850	0.837
85.876	80.058	1804.4	1821.0	1.005	1.230	0.845	0.832
84.834	78.786	1862.0	1831.9	1.006	1.224	0.844	0.831
84.821	78.774	1814.7	1832.0	1.006	1.224	0.844	0.831
81.356	74.694	1850.3	1866.8	1.010	1.204	0.841	0.828
78.703	71.710	1916.7	1892.3	1.013	1.190	0.839	0.826
77.008	69.865	1894.8	1908.0	1.015	1.182	0.838	0.824
75.184	67.923	1936.6	1924.5	1.017	1.173	0.836	0.823
75.094	67.825	1914.7	1925.3	1.017	1.172	0.836	0.823
72.393	65.040	1962.7	1948.7	1.021	1.159	0.834	0.820
68.922	61.583	1983.9	1977.3	1.027	1.144	0.832	0.818
66.725	59.460	2006.5	1994.6	1.031	1.134	0.830	0.816
65.342	58.148	1990.7	2005.1	1.034	1.128	0.830	0.815
64.300	57.172	2006.5	2012.8	1.036	1.124	0.829	0.814
63.148	56.098	2031.8	2021.3	1.038	1.120	0.828	0.814
57.550	51.033	2064.7	2059.8	1.051	1.099	0.825	0.810
56.131	49.784	2059.2	2068.9	1.055	1.094	0.824	0.809
53.170	47.205	2085.9	2087.0	1.063	1.084	0.823	0.807
51.854	46.069	2095.5	2094.8	1.067	1.080	0.822	0.807
47.000	41.940	2122.2	2121.5	1.082	1.065	0.820	0.804
44.069	39.480	2126.3	2136.1	1.091	1.057	0.818	0.803
41.974	37.728	2143.4	2145.9	1.099	1.052	0.818	0.802
41.431	37.277	2148.3	2148.4	1.101	1.050	0.817	0.802
37.452	33.963	2165.4	2165.2	1.116	1.041	0.816	0.800
34.610	31.598	2168.8	2176.0	1.127	1.035	0.815	0.799
33.071	30.311	2181.1	2181.4	1.134	1.032	0.815	0.799
31.524	29.018	2185.9	2186.5	1.141	1.029	0.814	0.798
30.977	28.559	2182.5	2188.2	1.143	1.028	0.814	0.798
27.918	25.981	2196.2	2197.1	1.157	1.023	0.813	0.797
22.982	21.757	2201.7	2208.7	1.182	1.015	0.812	0.796
21.191	20.201	2216.0	2212.0	1.192	1.013	0.812	0.796
20.620	19.702	2211.3	2212.7	1.195	1.012	0.812	0.796
17.627	17.054	2210.6	2217.2	1.211	1.009	0.812	0.795
16.989	16.482	2225.6	2217.9	1.215	1.008	0.812	0.795
13.081	12.924	2220.9	2220.8	1.239	1.005	0.812	0.795
12.205	12.110	2227.0	2221.1	1.244	1.004	0.812	0.795
10.386	10.401	2227.7	2221.3	1.256	1.003	0.812	0.795
9.593	9.647	2223.6	2221.2	1.261	1.003	0.812	0.795
7.204	7.339	2226.3	2220.3	1.277	1.002	0.812	0.795
5.418	5.576	2226.3	2218.8	1.290	1.001	0.812	0.795
3.881	4.031	2226.3	2217.1	1.301	1.000	0.812	0.795
1.386	1.462	2225.6	2213.3	1.319	1.000	0.812	0.796
0.000	0.000	2210.6	2210.6	1.330	1.000	0.813	0.796

(B) UNIQUAC

100.000	100.000	1655.8	1655.8	1.000	1.326	0.859	0.848
95.814	93.467	1701.0	1708.8	1.000	1.296	0.855	0.843
90.708	86.248	1759.3	1769.1	1.002	1.261	0.850	0.837
85.882	80.047	1804.4	1821.8	1.006	1.231	0.845	0.832
84.834	78.770	1862.0	1832.8	1.007	1.225	0.844	0.831
84.829	78.767	1814.7	1832.8	1.007	1.225	0.844	0.831
81.361	74.691	1850.3	1867.7	1.010	1.205	0.841	0.828
78.702	71.705	1916.7	1893.3	1.013	1.191	0.839	0.826
77.009	69.865	1894.8	1908.9	1.015	1.182	0.838	0.824
75.184	67.926	1936.6	1925.3	1.018	1.173	0.836	0.823
75.095	67.830	1914.7	1926.1	1.018	1.172	0.836	0.822
72.392	65.048	1962.7	1949.5	1.022	1.159	0.834	0.820
68.922	61.599	1983.9	1978.0	1.028	1.143	0.832	0.818
66.722	59.478	2006.5	1995.3	1.032	1.134	0.830	0.816
65.337	58.167	1990.7	2005.7	1.034	1.128	0.830	0.815
64.293	57.190	2006.5	2013.5	1.037	1.124	0.829	0.814
63.144	56.121	2031.8	2021.9	1.039	1.119	0.828	0.814
57.544	51.062	2064.7	2060.1	1.052	1.098	0.825	0.810
56.125	49.812	2059.2	2069.2	1.056	1.093	0.824	0.809
53.160	47.233	2085.9	2087.3	1.064	1.083	0.823	0.807
51.847	46.099	2095.5	2095.0	1.068	1.079	0.822	0.807
46.992	41.971	2122.2	2121.4	1.082	1.064	0.820	0.804
44.063	39.510	2126.3	2136.0	1.092	1.056	0.818	0.803
41.965	37.756	2143.4	2145.7	1.100	1.051	0.818	0.802
41.424	37.305	2148.3	2148.1	1.102	1.050	0.817	0.802
37.444	33.987	2165.4	2164.8	1.117	1.040	0.816	0.800
34.606	31.621	2168.8	2175.5	1.128	1.034	0.815	0.799
33.064	30.329	2181.1	2180.9	1.135	1.031	0.815	0.799
31.520	29.036	2185.9	2185.9	1.141	1.028	0.814	0.798
30.975	28.578	2182.5	2187.6	1.144	1.027	0.814	0.798
27.913	25.992	2196.2	2196.5	1.158	1.022	0.813	0.797
22.981	21.764	2201.7	2208.0	1.182	1.015	0.813	0.796
21.191	20.205	2216.0	2211.3	1.192	1.013	0.812	0.796
20.619	19.704	2211.3	2212.0	1.195	1.012	0.812	0.796
17.627	17.052	2210.6	2216.5	1.211	1.009	0.812	0.795
16.988	16.479	2225.6	2217.2	1.215	1.008	0.812	0.795
13.082	12.917	2220.9	2220.1	1.238	1.005	0.812	0.795
12.208	12.105	2227.0	2220.5	1.243	1.004	0.812	0.795
10.388	10.393	2227.7	2220.7	1.254	1.003	0.812	0.795
9.594	9.638	2223.6	2220.5	1.260	1.003	0.812	0.795
7.205	7.330	2226.3	2219.8	1.275	1.001	0.812	0.795
5.419	5.568	2226.3	2218.4	1.287	1.001	0.812	0.795
3.882	4.024	2226.3	2216.8	1.298	1.000	0.812	0.795
1.386	1.458	2225.6	2213.2	1.316	1.000	0.812	0.796
0.000	0.000	2210.6	2210.6	1.326	1.000	0.813	0.796

TABLE 14. VLE Results for MAM + DME at 0°C.

Mole Percent MAM		P (KPa)		Activity Coeff		Fugacity Coeff	
Liquid	Vapor	Meas	Calc	MAM	DME	MAM	DME
(A) NRTL							
100.000	100.000	134.2	134.2	1.000	1.525	0.976	0.974
95.520	88.155	146.0	145.9	1.001	1.481	0.974	0.971
90.449	77.395	158.9	158.2	1.003	1.433	0.972	0.969
83.514	65.815	174.1	173.4	1.010	1.372	0.969	0.966
79.017	59.682	182.1	182.4	1.016	1.334	0.967	0.964
73.094	52.790	193.2	193.3	1.028	1.288	0.965	0.962
63.975	44.060	208.2	208.2	1.052	1.224	0.963	0.959
58.924	39.920	215.1	215.6	1.070	1.191	0.961	0.958
56.417	38.003	219.0	219.0	1.080	1.176	0.961	0.957
49.158	32.849	228.2	228.3	1.115	1.136	0.959	0.955
44.495	29.772	233.5	233.7	1.141	1.112	0.958	0.954
39.106	26.367	239.2	239.5	1.177	1.088	0.957	0.953
34.201	23.352	243.7	244.3	1.215	1.068	0.956	0.952
27.522	19.284	250.0	250.4	1.276	1.045	0.955	0.951
23.612	16.875	253.2	253.6	1.318	1.033	0.954	0.950
18.578	13.692	257.4	257.5	1.378	1.021	0.954	0.950
13.385	10.250	261.8	261.1	1.451	1.011	0.953	0.949
9.807	7.741	266.3	263.3	1.508	1.006	0.953	0.948
3.826	3.203	266.8	266.4	1.617	1.001	0.952	0.948
2.164	1.845	266.9	267.2	1.651	1.000	0.952	0.948
0.000	0.000	268.0	268.0	1.697	1.000	0.952	0.948
(B) UNIQUAC							
100.000	100.000	134.2	134.2	1.000	1.527	0.976	0.974
95.520	88.151	146.0	145.9	1.001	1.482	0.974	0.971
90.450	77.392	158.9	158.2	1.003	1.433	0.972	0.969
83.514	65.817	174.1	173.4	1.010	1.372	0.969	0.966
79.017	59.687	182.1	182.4	1.016	1.334	0.967	0.964
73.094	52.797	193.2	193.3	1.028	1.288	0.965	0.962
63.975	44.067	208.2	208.2	1.052	1.223	0.963	0.959
58.924	39.925	215.1	215.6	1.070	1.191	0.961	0.958
56.417	38.007	219.0	219.0	1.080	1.176	0.961	0.957
49.158	32.850	228.2	228.3	1.115	1.135	0.959	0.955
44.495	29.771	233.5	233.7	1.141	1.112	0.958	0.954
39.106	26.363	239.2	239.5	1.177	1.088	0.957	0.953
34.201	23.346	243.7	244.3	1.215	1.068	0.956	0.952
27.522	19.277	250.0	250.4	1.276	1.045	0.955	0.951
23.612	16.869	253.2	253.7	1.317	1.034	0.954	0.950
18.578	13.688	257.4	257.5	1.378	1.021	0.954	0.950
13.385	10.248	261.8	261.1	1.451	1.011	0.953	0.949
9.807	7.742	266.3	263.3	1.508	1.006	0.953	0.948
3.826	3.205	266.8	266.4	1.618	1.001	0.952	0.948
2.164	1.847	266.9	267.2	1.652	1.000	0.952	0.948
0.000	0.000	268.0	268.0	1.700	1.000	0.952	0.948

TABLE 15. VLE Results for ANI + DME at 90°C.

Mole Percent ANI		P (KPa)		Activity Coeff		Fugacity Coeff	
Liquid	Vapor	Meas	Calc	ANI	DME	ANI	DME
(A) NRTL							
100.000	100.000	3.6	3.6	1.000	1.365	0.998	1.000
96.861	4.096	88.0	87.0	1.000	1.359	0.971	0.993
91.516	1.539	235.2	230.7	1.001	1.346	0.927	0.980
88.004	1.085	319.0	325.9	1.002	1.336	0.898	0.972
84.172	0.818	440.8	430.2	1.003	1.324	0.867	0.963
78.657	0.602	592.9	580.7	1.006	1.305	0.823	0.950
74.225	0.494	705.9	701.4	1.010	1.288	0.789	0.940
69.851	0.419	829.9	819.9	1.016	1.270	0.756	0.930
64.439	0.351	970.9	965.1	1.025	1.246	0.717	0.918
59.158	0.302	1110.7	1104.7	1.038	1.222	0.681	0.906
54.651	0.268	1231.9	1222.0	1.052	1.200	0.651	0.896
46.964	0.223	1436.0	1417.5	1.085	1.163	0.603	0.880
43.282	0.204	1516.1	1509.2	1.106	1.145	0.581	0.872
37.274	0.178	1688.7	1656.7	1.149	1.115	0.546	0.860
31.541	0.155	1828.4	1795.9	1.204	1.089	0.514	0.848
27.513	0.140	1890.7	1894.0	1.252	1.071	0.493	0.840
21.840	0.118	2046.2	2034.9	1.338	1.048	0.462	0.828
13.688	0.085	2278.4	2251.8	1.512	1.021	0.416	0.810
10.516	0.070	2305.6	2344.8	1.601	1.013	0.397	0.802
4.162	0.033	2564.8	2556.1	1.835	1.002	0.354	0.784
0.000	0.000	2722.9	2722.9	2.040	1.000	0.321	0.769
(B) UNIQUAC							
100.000	100.000	3.6	3.6	1.000	1.381	0.998	1.000
96.868	4.072	88.0	87.6	1.000	1.371	0.971	0.992
91.526	1.534	235.2	231.5	1.001	1.352	0.926	0.980
88.010	1.084	319.0	326.5	1.002	1.339	0.898	0.972
84.174	0.819	440.8	430.4	1.004	1.324	0.867	0.963
78.651	0.603	592.9	580.0	1.008	1.303	0.823	0.950
74.215	0.495	705.9	700.1	1.012	1.285	0.789	0.940
69.839	0.420	829.9	818.3	1.017	1.267	0.757	0.930
64.430	0.352	970.9	963.6	1.027	1.244	0.718	0.918
59.155	0.302	1110.7	1104.3	1.039	1.221	0.681	0.906
54.656	0.268	1231.9	1223.2	1.052	1.201	0.651	0.896
46.983	0.222	1436.0	1423.2	1.082	1.167	0.601	0.879
43.305	0.203	1516.1	1517.4	1.101	1.150	0.579	0.871
37.299	0.177	1688.7	1669.5	1.141	1.123	0.543	0.859
31.562	0.154	1828.4	1812.9	1.193	1.097	0.511	0.847
27.555	0.139	1890.7	1912.4	1.239	1.080	0.488	0.838
21.865	0.117	2046.2	2054.5	1.326	1.056	0.458	0.826
13.692	0.085	2278.4	2267.4	1.517	1.026	0.413	0.808
10.531	0.071	2305.6	2356.4	1.623	1.017	0.394	0.801
4.162	0.034	2564.8	2559.4	1.928	1.003	0.353	0.783
0.000	0.000	2722.9	2722.9	2.232	1.000	0.321	0.769

TABLE 16. VLE Results for ANI + DME at 40°C.

Mole Percent ANI		P (KPa)		Activity Coeff		Fugacity Coeff	
Liquid	Vapor	Meas	Calc	ANI	DME	ANI	DME
(A) NRTL							
100.000	100.000	0.2	0.2	1.000	1.202	1.000	1.000
95.427	0.436	39.9	42.5	1.000	1.198	0.980	0.994
90.806	0.211	85.5	85.4	1.000	1.192	0.959	0.989
86.935	0.145	122.1	121.3	1.001	1.186	0.943	0.984
82.751	0.107	163.1	160.2	1.002	1.180	0.925	0.979
75.088	0.070	231.7	231.1	1.005	1.165	0.892	0.969
71.530	0.060	263.9	263.8	1.008	1.157	0.878	0.965
65.160	0.046	318.8	321.7	1.014	1.142	0.852	0.957
60.162	0.039	355.8	366.5	1.021	1.130	0.833	0.951
53.510	0.031	426.1	425.1	1.033	1.112	0.807	0.944
46.596	0.025	482.6	484.7	1.051	1.093	0.782	0.936
42.346	0.022	524.0	520.7	1.066	1.081	0.767	0.931
36.970	0.018	568.1	565.8	1.088	1.066	0.748	0.925
29.551	0.014	634.3	627.2	1.130	1.046	0.723	0.917
25.751	0.012	669.5	658.7	1.157	1.037	0.711	0.913
20.416	0.010	717.1	703.1	1.204	1.025	0.693	0.907
17.415	0.008	728.8	728.5	1.235	1.019	0.683	0.903
14.237	0.007	771.5	755.9	1.272	1.013	0.672	0.900
10.792	0.005	799.8	786.4	1.319	1.008	0.661	0.896
4.644	0.002	868.7	843.7	1.423	1.002	0.639	0.888
0.000	0.000	890.8	890.8	1.523	1.000	0.621	0.882
(B) UNIQUAC							
100.000	100.000	0.2	0.2	1.000	1.211	1.000	1.000
95.430	0.434	39.9	42.6	1.000	1.203	0.980	0.994
90.809	0.211	85.5	85.5	1.001	1.195	0.959	0.989
86.937	0.145	122.1	121.4	1.001	1.187	0.943	0.984
82.750	0.107	163.1	160.2	1.003	1.179	0.925	0.979
75.084	0.070	231.7	230.9	1.006	1.164	0.893	0.969
71.525	0.060	263.9	263.5	1.009	1.156	0.878	0.965
65.157	0.046	318.8	321.5	1.015	1.142	0.852	0.957
60.163	0.039	355.8	366.5	1.021	1.130	0.833	0.951
53.516	0.031	426.1	425.7	1.032	1.114	0.807	0.943
46.605	0.025	482.6	486.2	1.049	1.096	0.782	0.935
42.357	0.022	524.0	522.9	1.062	1.085	0.766	0.931
36.981	0.018	568.1	568.7	1.083	1.071	0.747	0.925
29.560	0.014	634.3	630.9	1.123	1.052	0.722	0.916
25.758	0.012	669.5	662.5	1.150	1.043	0.709	0.912
20.420	0.010	717.1	706.9	1.199	1.030	0.692	0.906
17.422	0.008	728.8	731.9	1.233	1.023	0.682	0.903
14.238	0.007	771.5	758.8	1.276	1.017	0.671	0.899
10.793	0.005	799.8	788.5	1.332	1.010	0.660	0.896
4.644	0.002	868.7	844.4	1.468	1.002	0.638	0.888
0.000	0.000	890.8	890.8	1.616	1.000	0.621	0.882

TABLE 17. Activity Coefficient Model Parameters.

COMPONENTS	T (°C)	NRTL PARAMETERS			UNIQUAC PARAMETERS	
		α	A_{12} (K)	A_{21} (K)	A_{12} (K)	A_{21} (K)
PHE=1, DME=2	80	0.2	601.61	-781.02	389.26	-389.40
PHE=1, DME=2	46	0.2	591.19	-800.05	389.44	-389.05
PHE=1, PAM=2	160	-.-	---.--	----.--	---.--	---.--
PHE=1, PAM=2	80	-.-	---.--	----.--	---.--	---.--
EFO=1, DME=2	80	0.2	-657.26	1001.57	-299.89	519.02
EFO=1, DME=2	0	0.2	-349.71	466.38	-154.11	224.15
EBE=1, PAM=2	160	0.2	-383.20	513.96	-14.49	14.49
EBE=1, PAM=2	80	0.2	-613.69	961.10	---.--	---.--
MAM=1, DME=2	80	0.2	30.72	70.44	-37.72	110.47
MAM=1, DME=2	0	0.2	-73.07	221.58	-60.39	166.16
ANI=1, DME=2	90	0.2	-365.52	705.96	-125.96	234.12
ANI=1, DME=2	40	0.2	-311.35	511.60	-116.46	177.16

COMPONENTS	T (°C)	RK PARAMETERS[*]			
		A	B	C	D
PHE=1, DME=2	80	-1.3199	-0.2862	0.1486	0.1165
PHE=1, DME=2	46	-1.7606	-0.5051	0.3229	0.0642
PHE=1, PAM=2	160	-2.6034	-0.5243	0.6057	0.3990
PHE=1, PAM=2	80	-5.4341	-1.4077	1.5666	1.1540

[*] $g^E/RT = x_1 x_2 [A + B(x_1-x_2) + C(x_1-x_2)^2 + D(x_1-x_2)^3]$

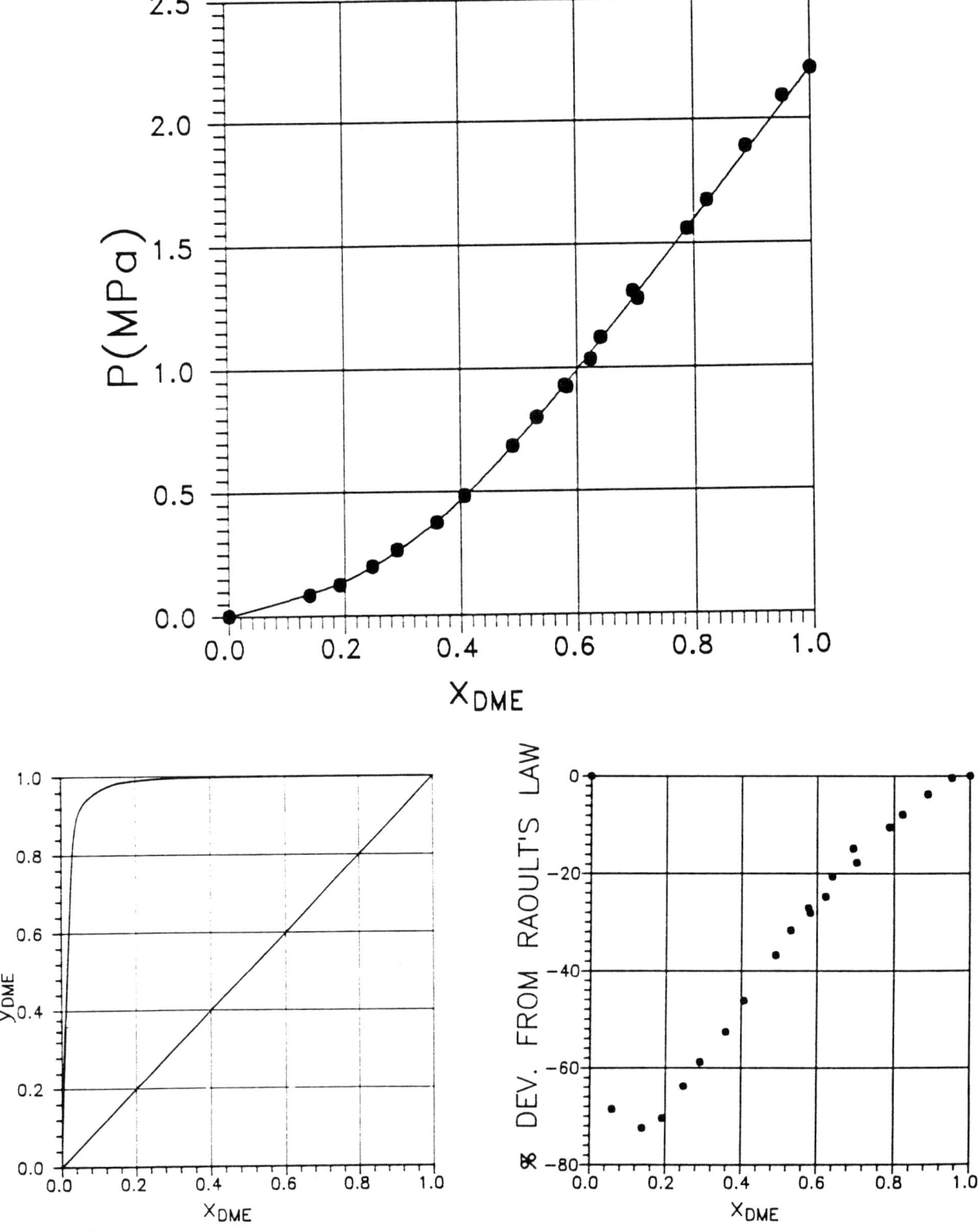

Figure 6. PHE + DME mixtures at 80°C: Experimental (•) and calculated (———) total pressures, (Top); equilibrium compositions, (Lower Left); and % deviations from Raoult's Law.

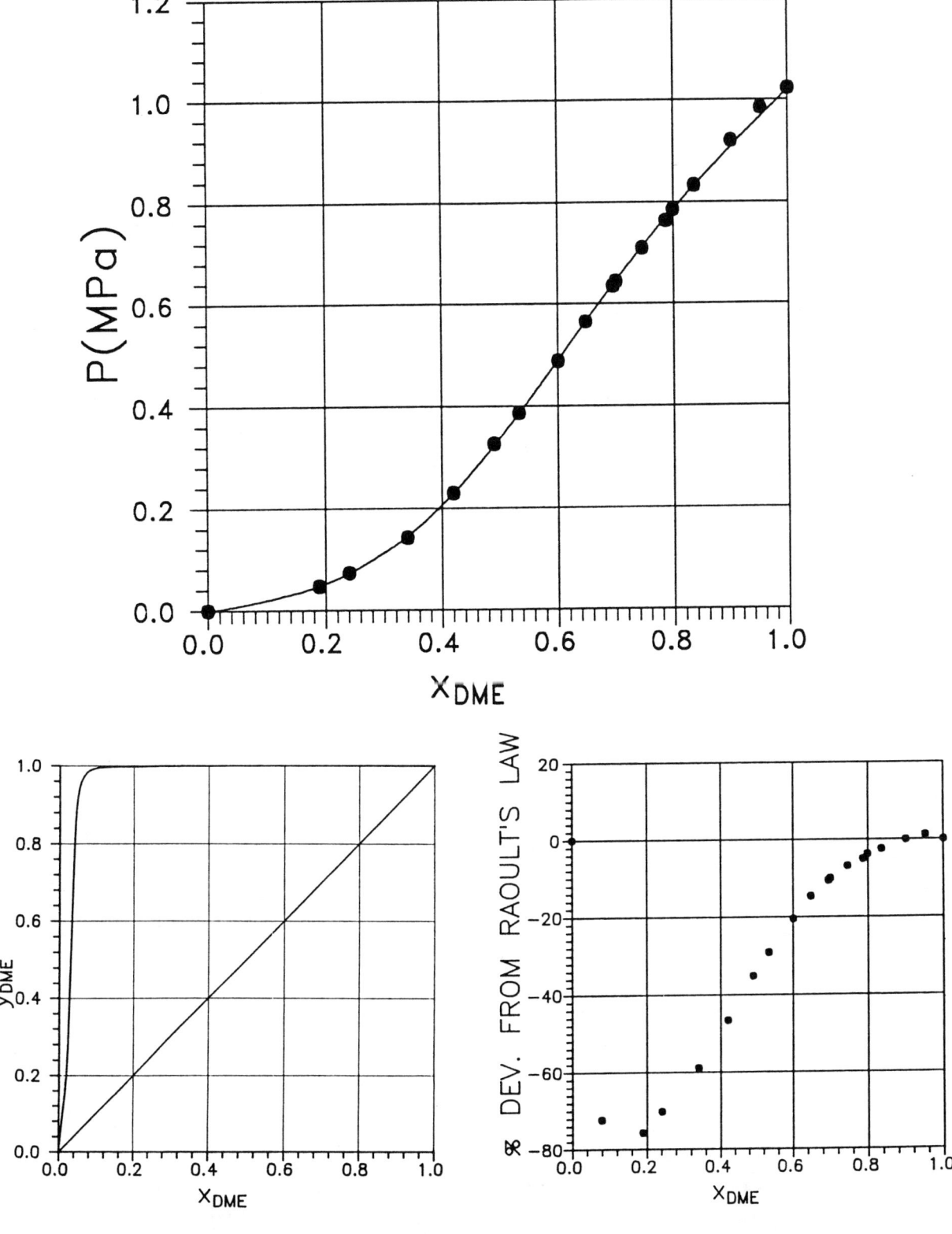

Figure 7. PHE + DME mixtures at 46°C: Experimental (•) and calculated (———) total pressures, (Top); equilibrium compositions, (Lower Left); and % deviations from Raoult's Law.

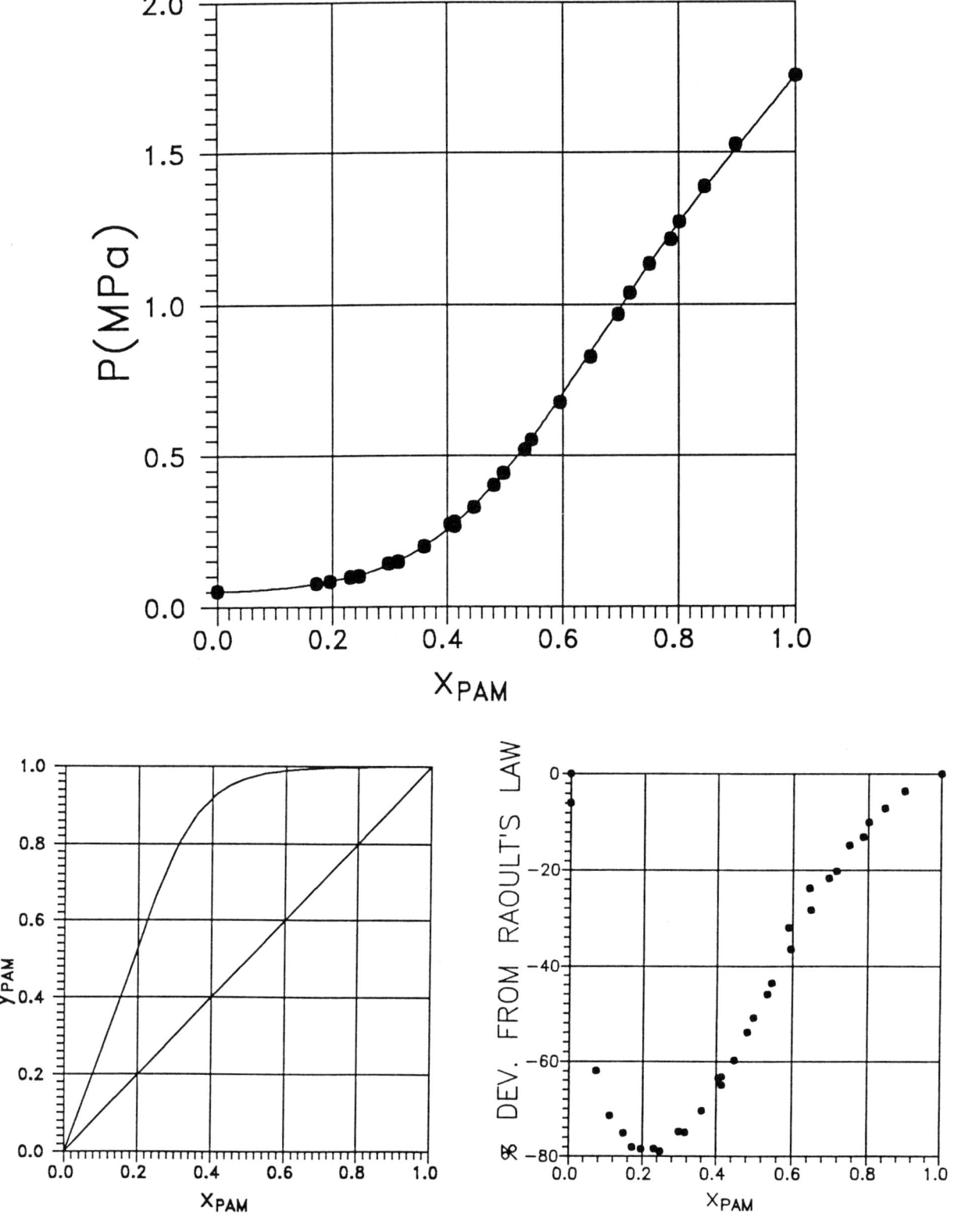

Figure 8. PHE + PAM mixtures at 160°C: Experimental (•) and calculated (———) total pressures, (Top); equilibrium compositions, (Lower Left); and % deviations from Raoult's Law.

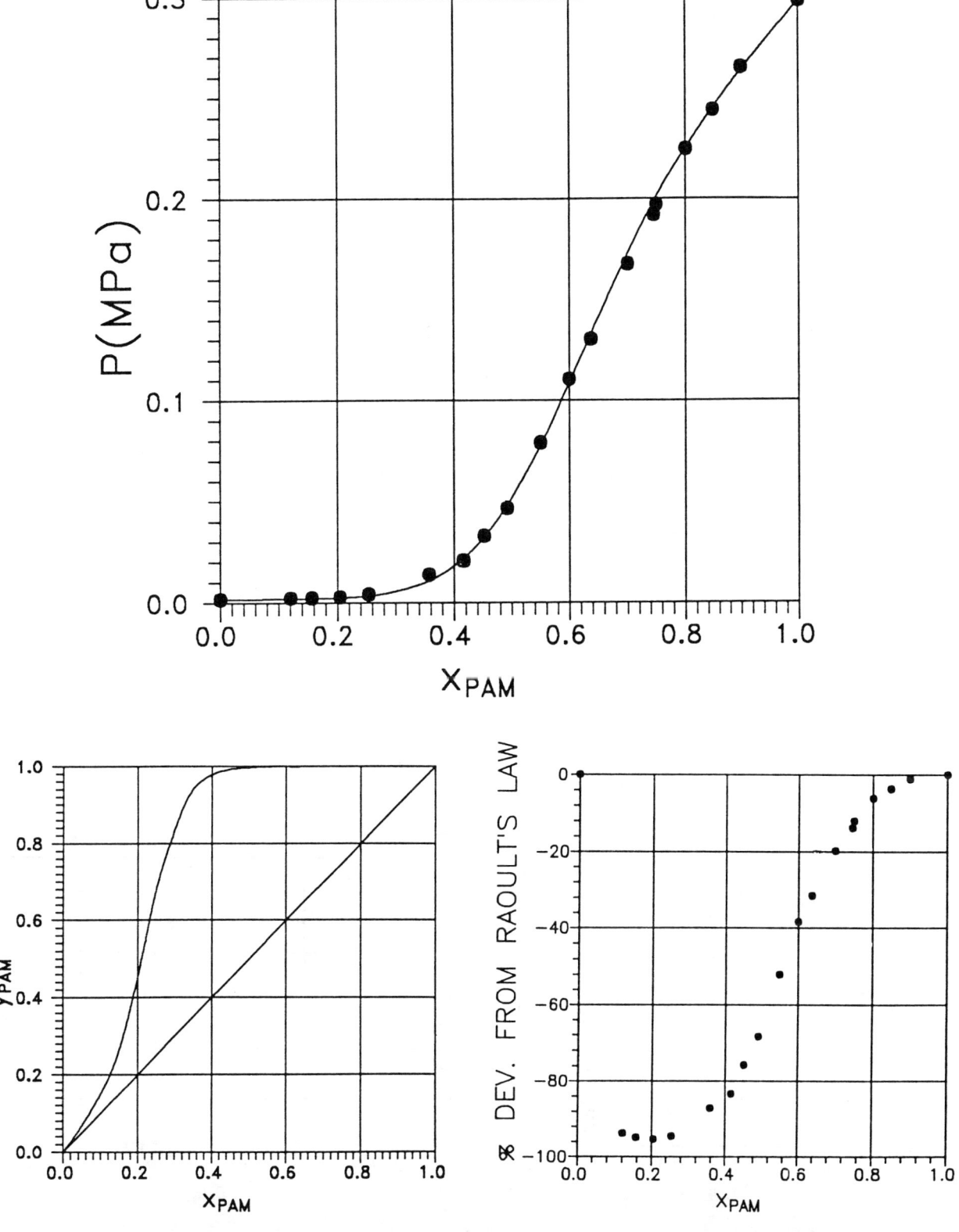

Figure 9. PHE + PAM mixtures at 80°C: Experimental (•) and calculated (———) total pressures, (Top); equilibrium compositions, (Lower Left); and % deviations from Raoult's Law.

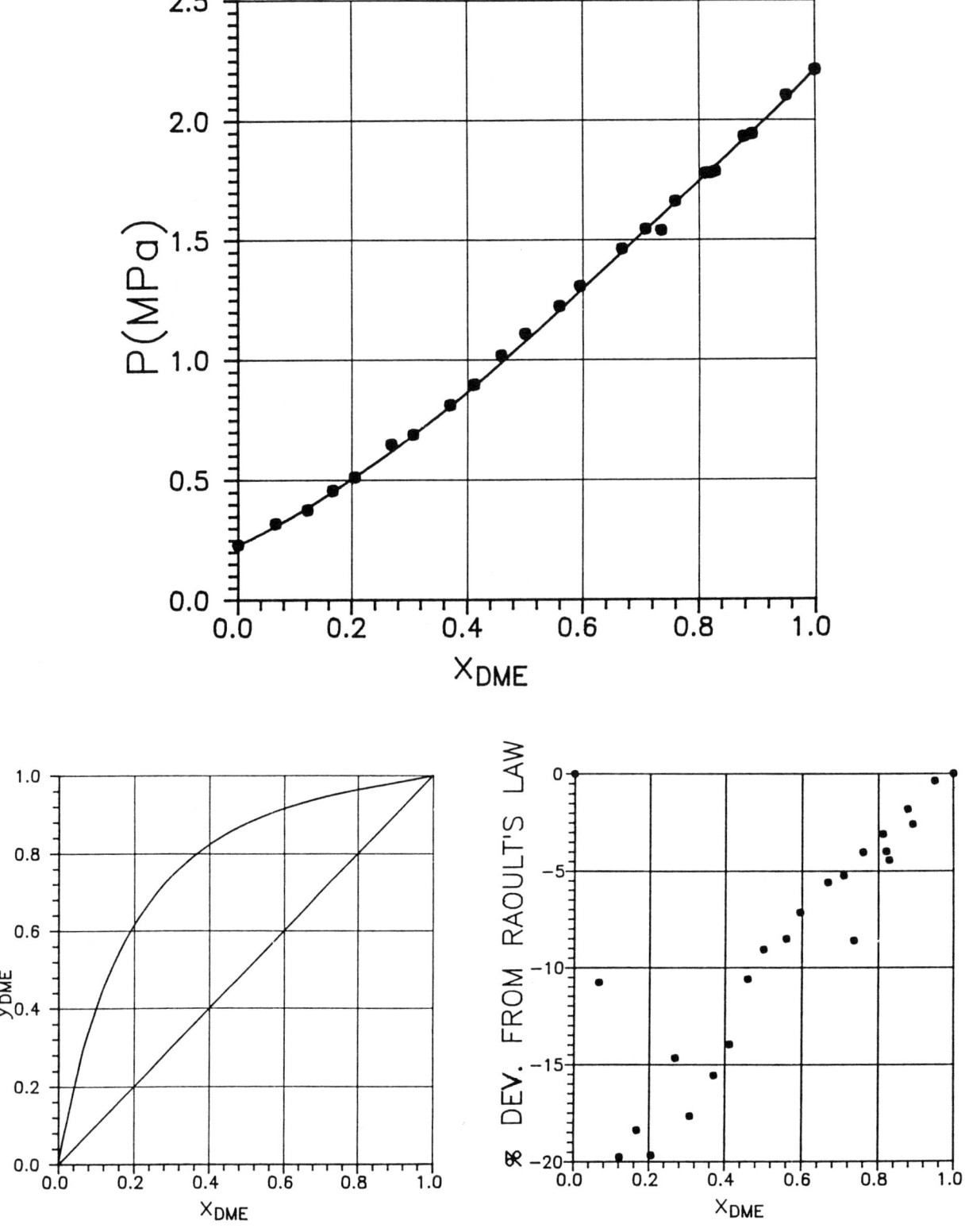

Figure 10. EFO + DME mixtures at 80°C: Experimental (•) and calculated (———) total pressures, (Top); equilibrium compositions, (Lower Left); and % deviations from Raoult's Law.

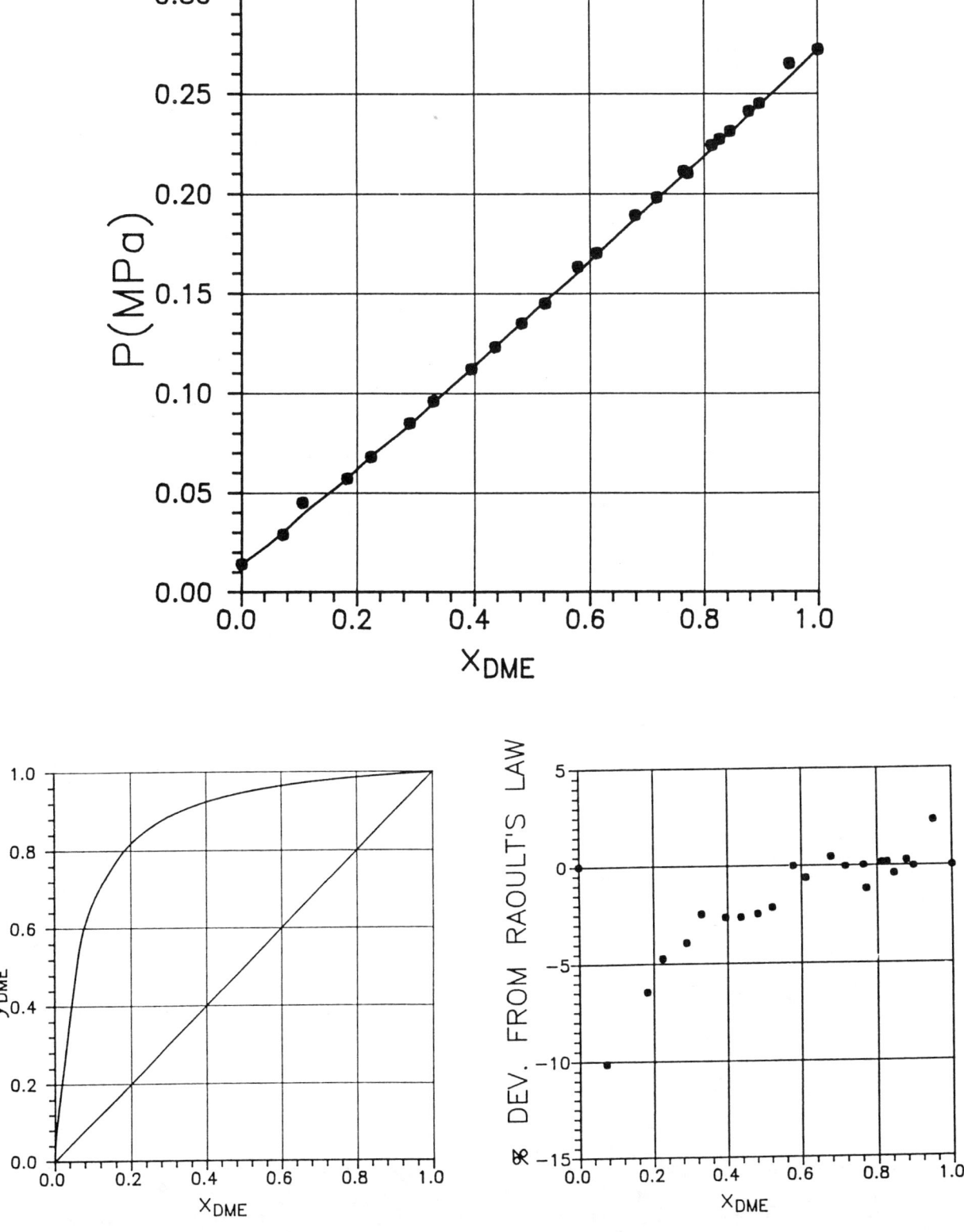

Figure 11. EFO + DME mixtures at 0°C: Experimental (•) and calculated (——) total pressures, (Top); equilibrium compositions, (Lower Left); and % deviations from Raoult's Law.

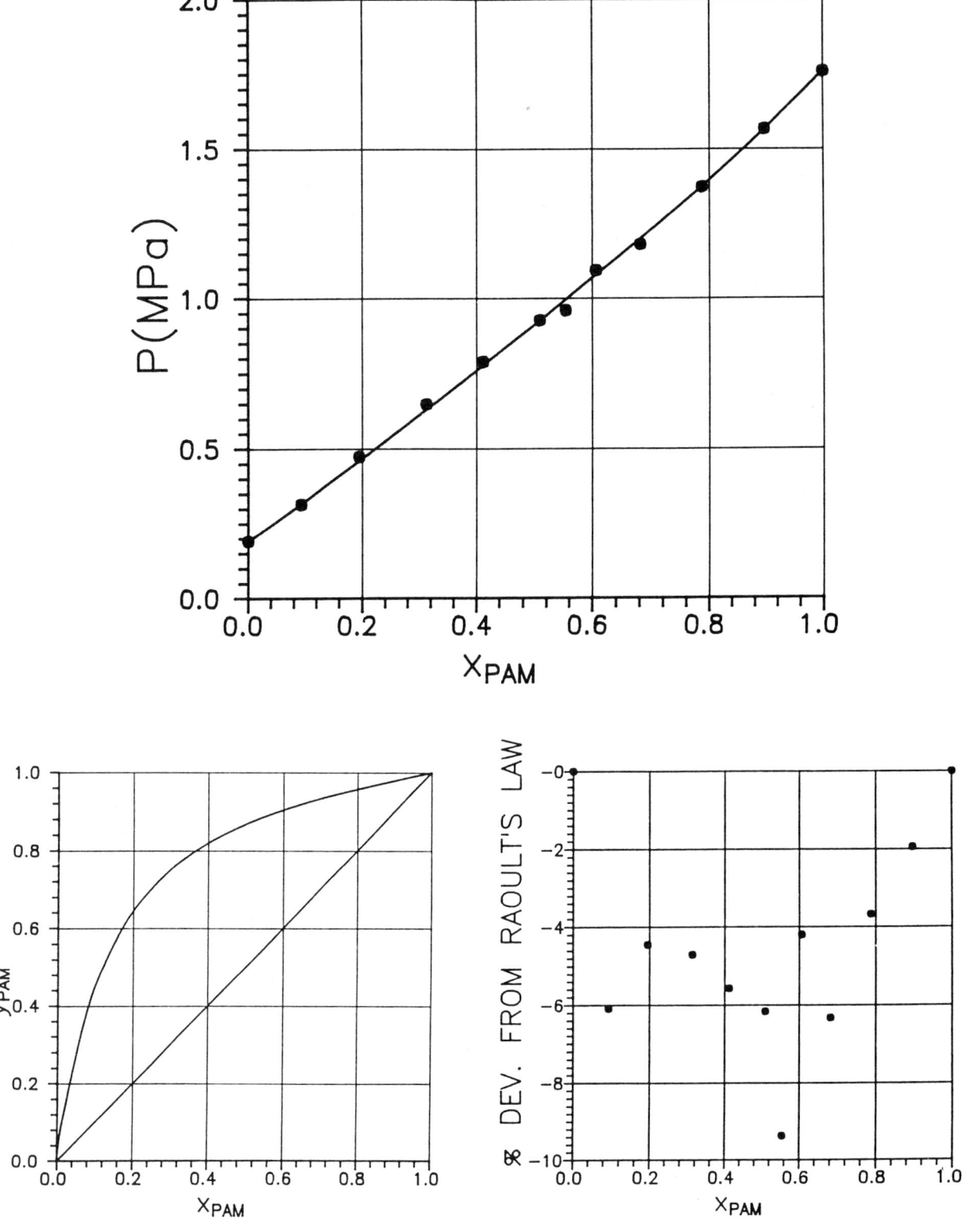

Figure 12. EBE + PAM mixtures at 160°C: Experimental (•) and calculated (———) total pressures, (Top); equilibrium compositions, (Lower Left); and % deviations from Raoult's Law.

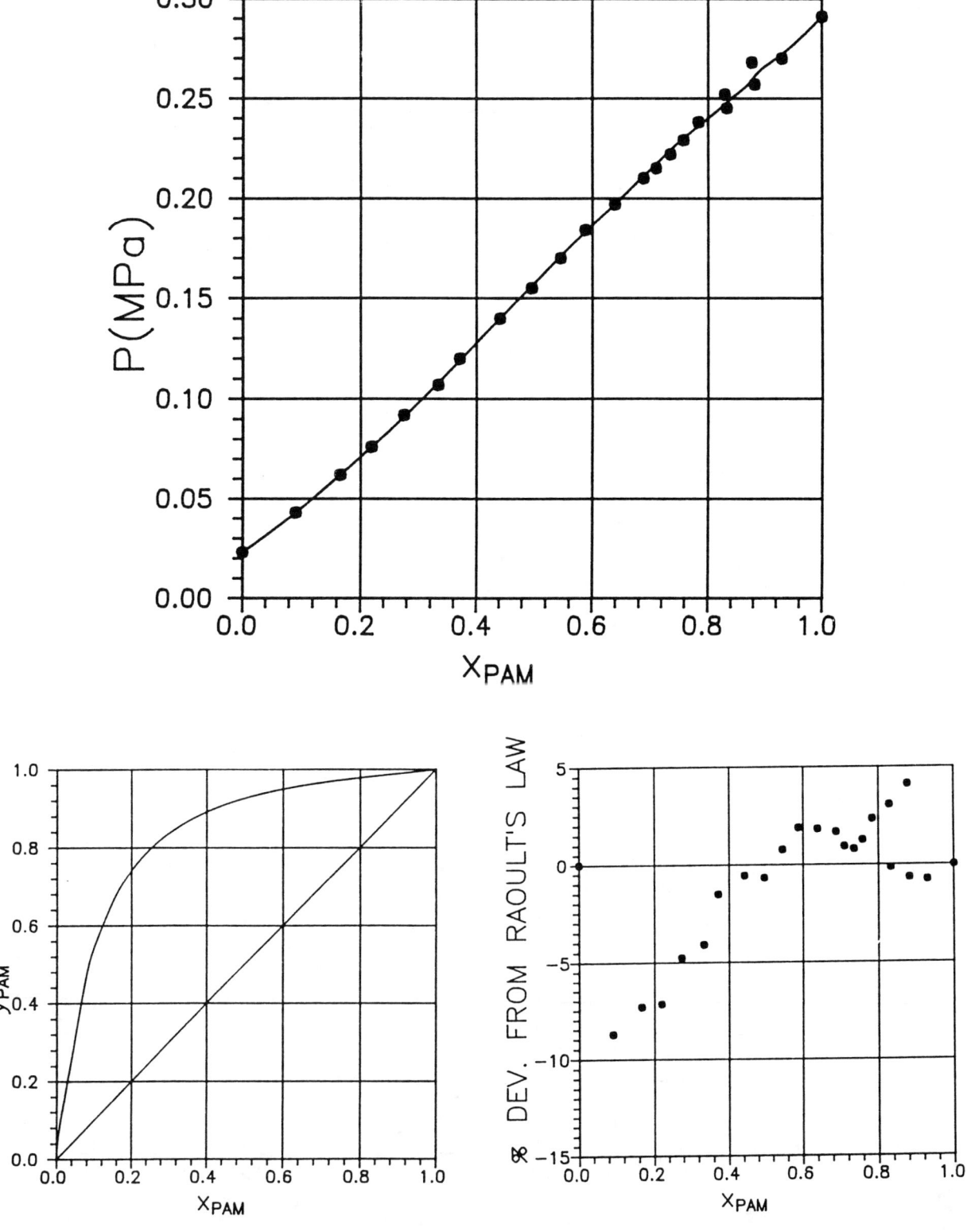

Figure 13. EBE + PAM mixtures at 80°C: Experimental (•) and calculated (———) total pressures, (Top); equilibrium compositions, (Lower Left); and % deviations from Raoult's Law.

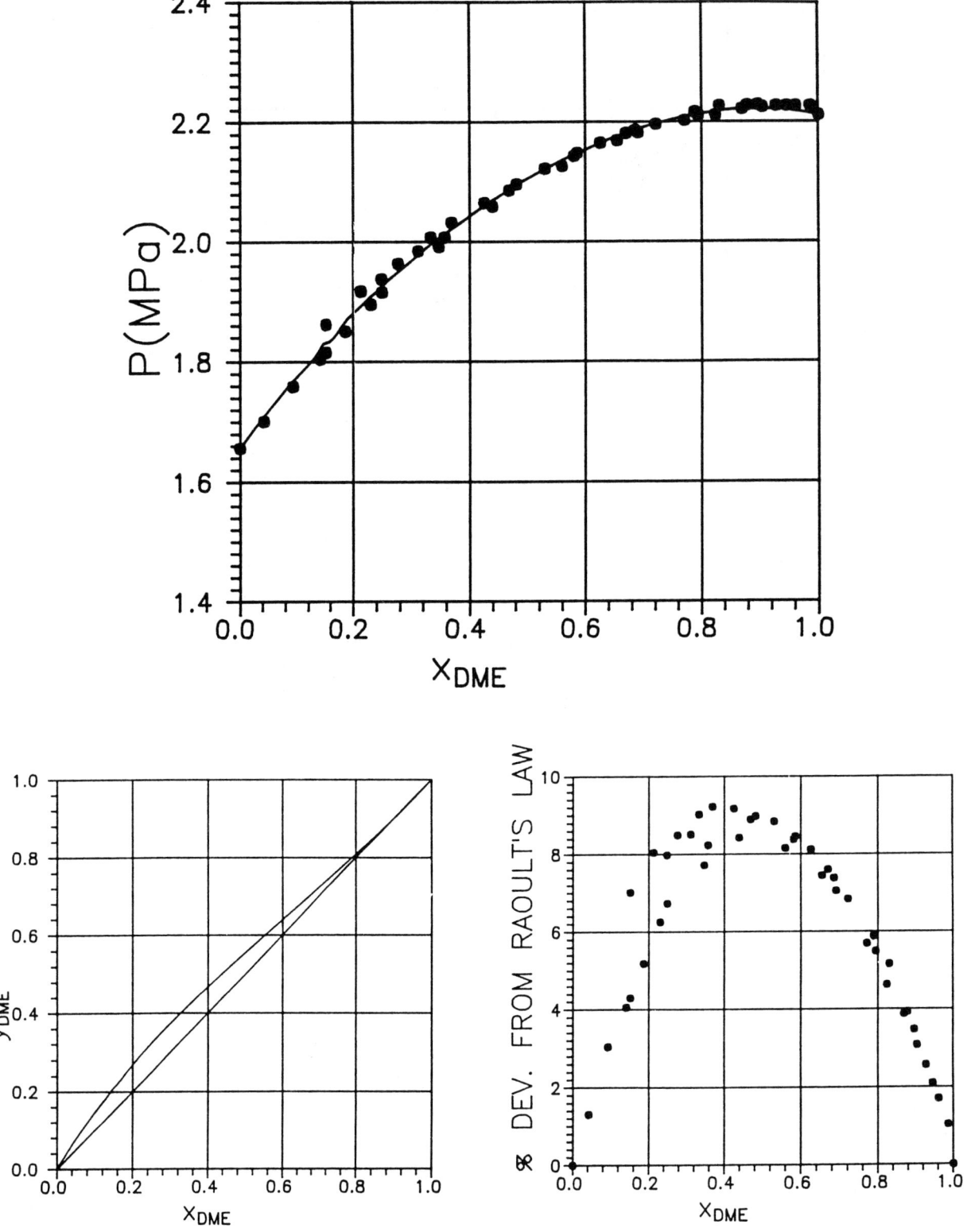

Figure 14. MAM + DME mixtures at 80°C: Experimental (•) and calculated (——) total pressures, (Top); equilibrium compositions, (Lower Left); and % deviations from Raoult's Law.

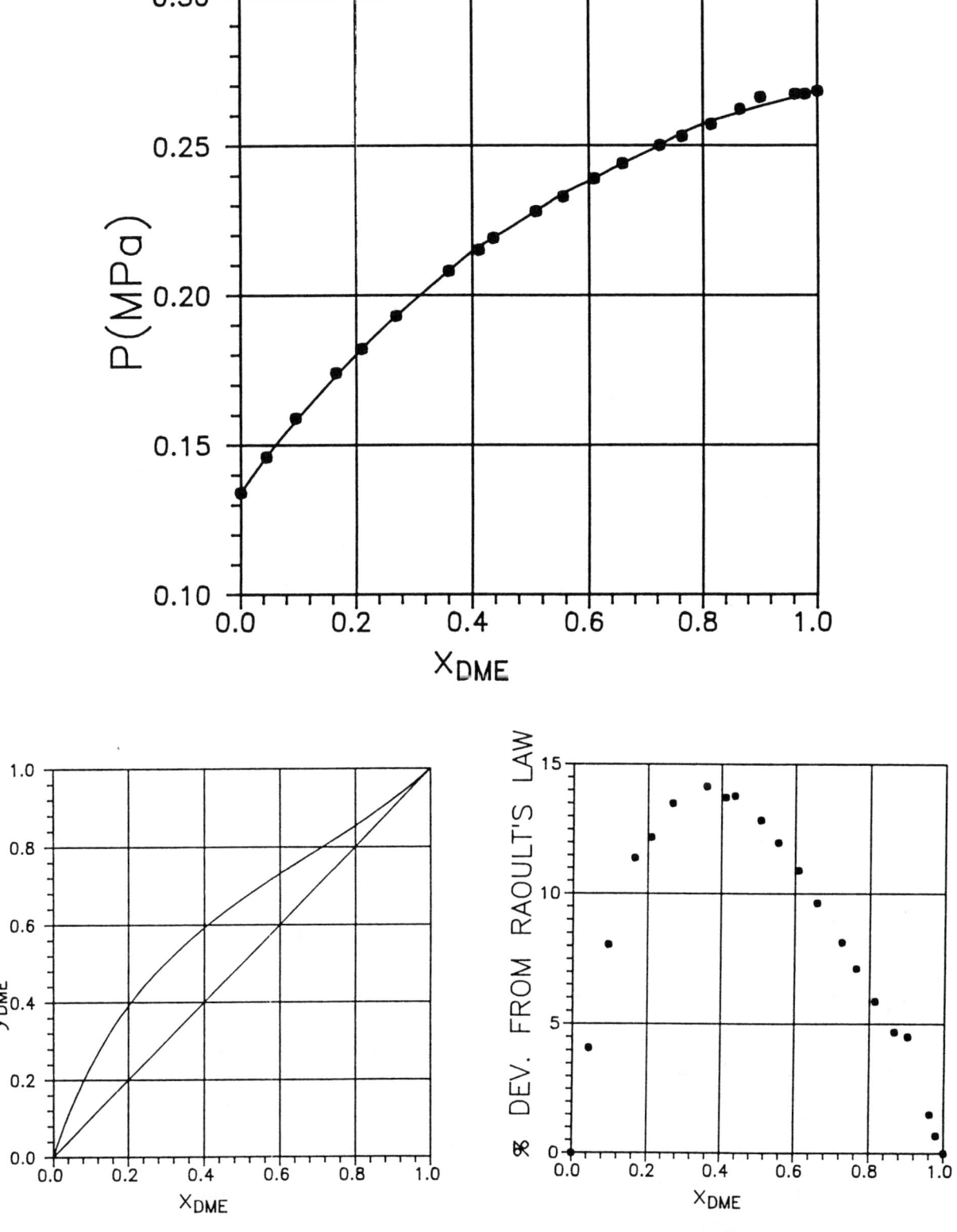

Figure 15. MAM + DME mixtures at 0°C: Experimental (•) and calculated (———) total pressures, (Top); equilibrium compositions, (Lower Left); and % deviations from Raoult's Law.

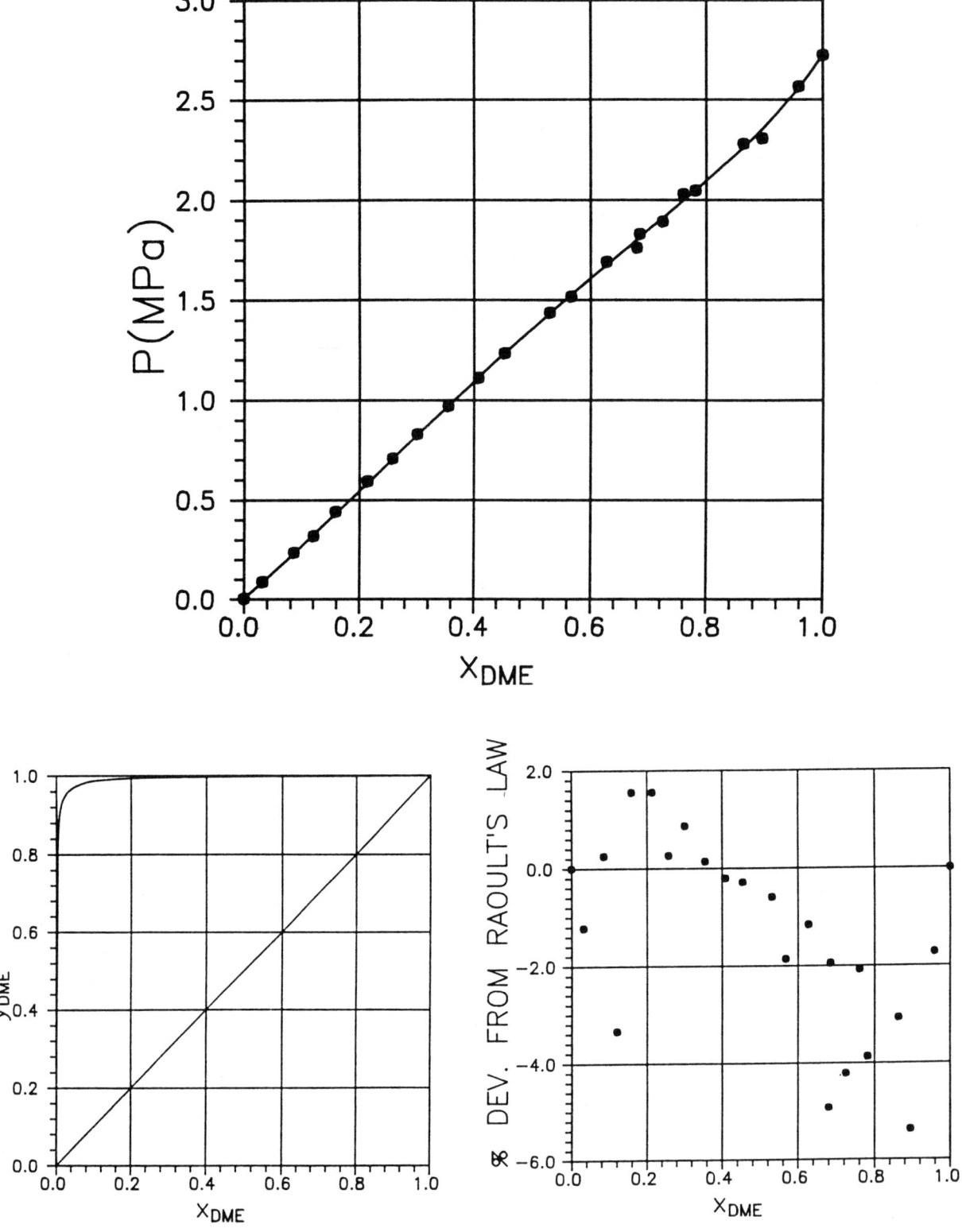

Figure 16. ANI + DME mixtures at 90°C: Experimental (•) and calculated (——) total pressures, (Top); equilibrium compositions, (Lower Left); and % deviations from Raoult's Law.

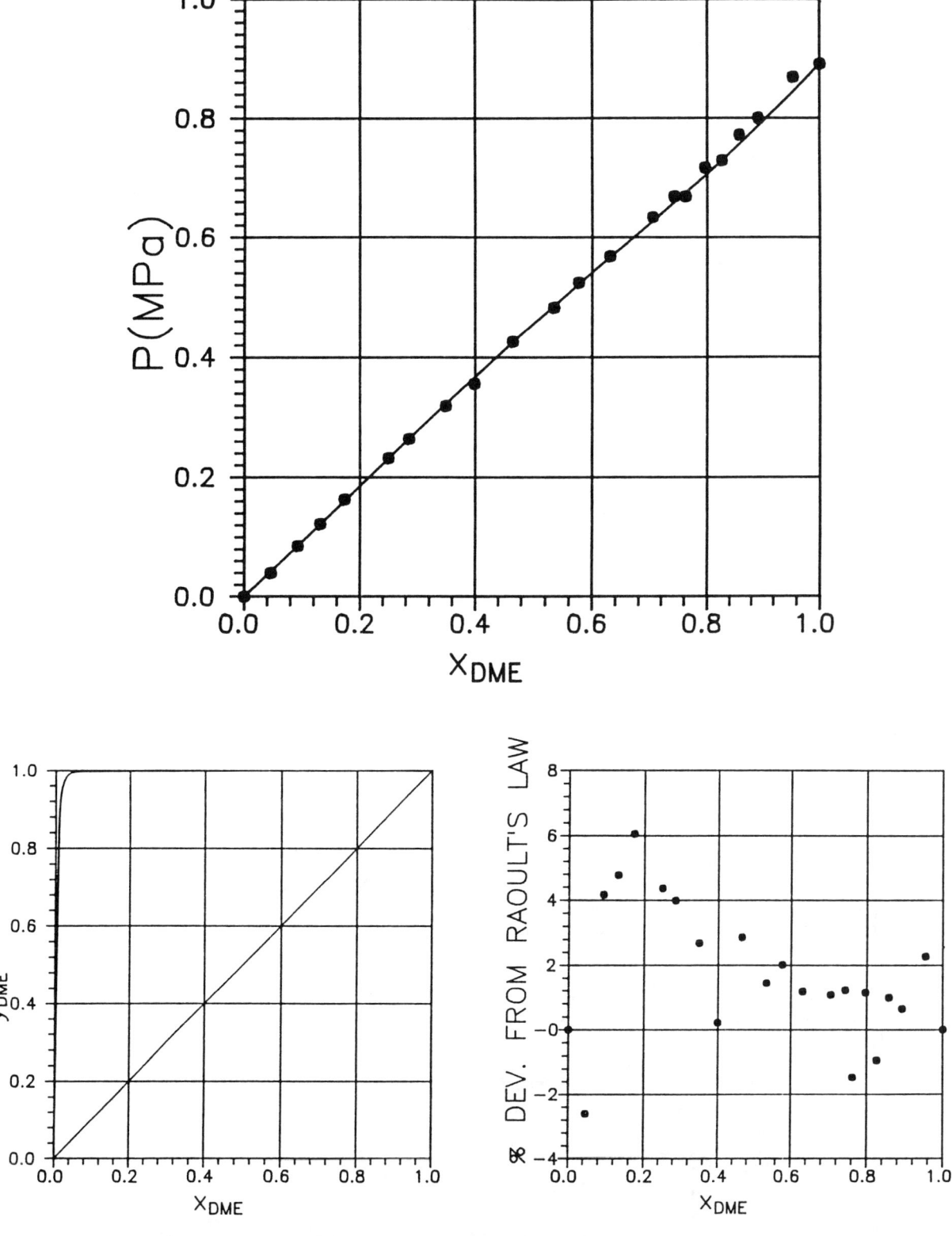

Figure 17. ANI + DME mixtures at 40°C: Experimental (•) and calculated (——) total pressures, (Top); equilibrium compositions, (Lower Left); and % deviations from Raoult's Law.

VAPOR PRESSURE OF 13 PURE INDUSTRIAL CHEMICALS

Thomas E. Daubert ■ The Pennsylvania State University,
Department of Chemical Engineering, University Park, PA 16801

Vapor pressures were measured from pressures of about 1 kPa to the decomposition point of each compound where possible. An ebulliometer and a capillary apparatus were used for subatmospheric and superatmospheric pressure measurements, respectively. This paper gives a complete listing of all new experimental data, compares the data with literature data where available, and smooths the data using a standard vapor pressure equation. Tests on the reliability of the data and tendencies to decompose or polymerize are discussed. Coefficients for the regressions of the experimental data are given together with the fitting accuracy for each compound studied.

The measurement of liquid vapor pressure of pure compounds from a pressure of about 1.33 kPa (100 mm Hg) to the critical point (when feasible) has been carried out in this laboratory since 1982 under the sponsorship of the Design Institute for Physical Property Data . Data taken for the 1982, 1983, and 1984 project years were reported by Daubert, Jalowka, and Goren ([1]). A paper describing the project's experimental and data analysis methods was published by Daubert and Jones ([2]). Data taken for project years 1985, 1986, and 1987, were published by Daubert and Hutchison ([3]). This paper reports on data for project years 1988 and 1989.

EXPERIMENTAL APPARATUS AND PROCEDURES

Two apparatuses are used to determine vapor pressures - an ebulliometer for use from 1 kPa to the prevailing atmospheric pressure and a capillary apparatus from atmospheric pressure to the critical point or decomposition. Both experimental setups are tested at least once a year by running isopropyl alcohol, for which excellent data are available, to check operation and accuracy of temperature and pressure measurements. Full descriptions of the apparatus were given by Daubert, Jalowka, and Goren ([1]) and updated by Daubert and Hutchison ([3]).

SOURCES OF CHEMICALS

	Compound/Source	Estimated Purity
1.	sec-Butyl acetate Aldrich	99%
2.	t-Butyl acetate Aldrich	99+%
3.	Methylacetoacetate Aldrich	99+%
4.	Ethyl-t-butyl ether Phillips	99+%
5.	2-Phenylpropionaldehyde Aldrich	98%
6.	2-Methylbenzofuran Aldrich	96%
7.	Diacetone alcohol Aldrich	98%
8.	Ethylthioethanol Aldrich	97%

9.	n-Butylamine Aldrich	99 + %
10.	Benzylamine Aldrich	99 + %
11.	N-Aminoethylethanolamine Union Carbide	No purity given
12.	Methane sulfonyl chloride Pennwalt	Manuf. purity
13.	1,2-Ethanedithiol Phillips	No purity given

MEASURED RESULTS

Table 1 summarizes the compounds studied giving the range of temperature and pressure with notes concerning the reasons for the limited range of experiments. Table 2 lists all raw experimental data in units of °C and pascals. Each experiment for each compound is listed separately with different 995X and 997X references indicating experiments in the low-pressure apparatus and 998X references indicating experiments in the high-pressure apparatus. Data smoothing was carried out as discussed in the next section.

METHODS OF DATA ANALYSIS

Data from the project are evaluated both qualitatively and quantitatively. Qualitative analyses are based on a plot of the experimental data on $\ln P$ vs. $1/T$ coordinates. As well as a rapid check on the accuracy of the measurements, this analysis shows whether the sample is decomposing or polymerizing as the temperature increases. If the slope of the vapor pressure curve on $\ln P - 1/T$ coordinates increases substantially as temperature increases, decomposition is indicated. Conversely a decrease in slope as temperature increases indicates polymerization. Any available experimental data from the literature are also included in the plot for comparison.

Quantitative analyses include fitting of the data to a regression equation to smooth the data and then carrying out several tests and comparisons between experimental and calculated points.

1. The data are fit to the modified Frost-Kalkwarf equation used by the DIPPR Compilation Project.

$$P = \exp(A + B/T + C \ln T + DT^E)$$

The exponent E is set to 2 or 6 according to which gives the most accurate fit with the coefficients determined by a least squares regression.

Two fits are carried out - one constrained at the critical point (either experimental or predicted) and another unconstrained and fit only over the data range. The constrained fit is carried out such that the thermodynamically correct slight S-shaped curve between the critical and normal boiling points is obtained which requires that the sign of coefficients A and D are positive while the sign of coefficients B and C are negative. The unconstrained fit does not assure that this "sign rule" will be passed although the experimental data may be accurate. For many compounds data above the start of decomposition or polymerization could cause the shape of the curve to be predicted inaccurately. Passing of the sign rule only signifies that the data may be accurate. Coefficients for fits of the data are in Table 3.

2. The constrained fit is tested by using the test described by Ambrose (3) which dictates that on a plot of $\Delta H/Z$ (equal to $RT^2 d(\ln P)/dT$) vs. reduced temperature, a minimum must exist in the 0.8-0.9 reduced temperature range.

3. If a line on a $\ln P$ vs $1/T$ plot is drawn between the melting point and the critical point all data must coincide or deviate positively from the line. This so called "line rule" plot is a quick but insensitive test of the data.

4. For both the constrained and unconstrained regressions, plots are made of ($\ln P_{calc} - \ln P_{exp}$) labeled (Delta $\ln P$) vs $1/T$ for each experimental data point and the critical point even if estimated in the case of the constrained fit. Individual point and overall deviations are tabulated. Such plots should indicate no substantial bias deviation. However, the constrained fit will sometimes show a bias when a predicted critical is used, thus questioning its validity. The magnification of deviations on this type plot is useful in both validating the data and the regression line.

Typical results of each of these tests were described by Daubert and Jones (2). This paper will present quantitative results only for regressions as presented in Table 3 and discussed below for each compound.

ANALYSIS BY COMPOUND

1. **sec-Butyl acetate** - The experimental data taken on this compound agreed well with the two points of literature data available. Figure 1 shows the experimental data for the low (9971) and high (9981 and 9982) pressure experiments together with the two points of experimental data (3 and 39) and the estimated critical point (380). The constrained regression line is also shown. Sets of coefficients for the constrained and unconstrained regressions with average errors are given in Table 3. As usual the unconstrained regression gives the better fit over the experimental data range but is not extrapolatable.

2. **t-Butyl acetate** - Figure 2 shows the experimental data for the low (9971) and high (998X) experiments and the unconstrained regression line. Only the normal boiling point (3) was available in the literature and is in total agreement with the data taken. This compound although stable showed somewhat erratic behavior at higher pressures. Extrapolation to the critical point (380) was reasonable. Table 3 gives both constrained and unconstrained regression coefficients.

3. **Methylacetoacetate** - Measured experimental data for the low (9971) and high (998X) pressure and the constrained regression line are shown in Figure 3. The data matched the literature normal boiling point (39) and extrapolated well to two low pressure literature points (3092 and 3111). However, the compound definitely began to decompose slightly below the highest temperature point (9982) at 229°C. The prediction of the critical pressure (380) was obviously low although the normal boiling point was measured accurately. Thus, the error of the constrained regression shown in Table 3 was high. The unconstrained regression should be used in this case but definitely not above 220°C.

4. **Ethyl-t-butyl ether** - Figure 4 is a plot of the low (9971) and high (998X) pressure experimental data and the unconstrained regression line. Steady boiling was impossible to maintain at the lowest pressures of experiment 9971 although the data appear reasonable. Decomposition began at approximately 200°C as shown by the two highest pressure points. A predicted critical pressure (380) was slightly low affecting the error in the constrained regression line shown in Table 3. Again the unconstrained regression should be used to an upper limit of 200°C.

5. **2-Phenylpropionaldehyde** - The experimental low (9971) and high (9981 and 9982) pressure data and the constrained regression line are shown in Figure 5. The highest temperature (238°C) point is good, but rapid decomposition began slightly above this point. Scatter in the high pressure data near the normal boiling point could not be eliminated although three experiments were run. The normal boiling point extrapolated from the low pressure data (380) and the predicted critical (380) agree very well with the data. Table 3 shows essentially equal fits whether constrained or unconstrained.

6. **2-Methylbenzofuran** - Experimental data (9971, 9981, and 9982), literature data from four sources (2867, 4284, 4286, and 4340), and the constrained correlation line shown in Figure 6 are in excellent agreement over the entire range of data. The predicted critical (380) fits the data well. Table 3 shows both regressions fit the data well with the constrained regression extrapolatable to the critical point.

7. **Diacetone alcohol** - Figure 7 shows the low (9972) and high (9986 and 9989) pressure data together with a predicted regression line and one low temperature experimental data point (39). Decomposition occurs above about 145°C. This qualitatively agrees with Fuge et al. (1391) who state that decomposition to mesityl oxide and water begins at about 130°C with a normal boiling point of 168°C. Thus, all data above 145°C should be disregarded. The constrained regression of Table 3 are applicable coefficients and usable below this temperature.

8. **Ethylthioethanol** - The experimental data and the unconstrained regression line are shown in Figure 8. No literature data were available for this compound. Prediction of the critical (380) gives a pressure considerably lower than the data extrapolation. Table 3 shows the regression of the data with the unconstrained regression fitting the data well but not passing the sign rule, thus being unable to be extrapolated.

9. **n-Butylamine** - Figure 9 shows experimental data in both ranges as well as the few points of literature data (3, 1222, and 1946) are in good agreement. The constrained regression line fits the data well including the predicted critical (380). Both regressions in Table 3 are usable and pass the sign rule.

10. **Benzylamine** - Low and high pressure experimental data and the constrained regression line are shown on Figure 10 together with a predicted critical point (380) and one point of literature data (2668). With the exception of the highest temperature experiment (9982) at 266°C where the compound was decomposing, the data are consistent and extrapolate well to the predicted critical. Table 3 shows both constrained and unconstrained regressions to be valid.

11. **N-Aminoethyl ethanolamine** - Figure 11 shows both high and low pressure data and the constrained regression line. No literature data were available. The predicted critical pressure is low if compared with an extrapolated data line. Table 3 shows a reasonable fit for all data for the constrained regression.

12. **Methane sulfonyl chloride** - High pressure data for this compound are unavailable as reaction with mercury occurs in the sample tube. As no literature data exist, Figure 12 shows only the low pressure data, the predicted critical (380), and the constrained regression line which extrapolates well to the predicted critical. Both the constrained and unconstrained regressions in Table 3 correlate the data accurately.

13. **1,2-Ethanedithiol** - Because of limited sample availability only low pressure data together with an extrapolated normal boiling point (380), a predicted critical (380), and the constrained regression line are shown in Figure 13. No literature data exist. The data are accurately represented by both the constrained and unconstrained regressions of Table 3.

ACKNOWLEDGEMENTS

Financial support of the Design Institute for Physical Property Data Pure Component Vapor Pressure Project is gratefully acknowledged. Experimental work of graduate and undergraduate students during the 1988-1989 project period especially Gary Hutchison and Daniel Blessner forms the basis of this paper.

LITERATURE CITED

1. Daubert, T. E., Jalowka, J. W., Goren, V., AIChE Symposium Series, 83 (256), 128 (1987).

2. Daubert, T. E., Jones, D. K., AIChE Symposium Series, 86 (275), 29 (1990).

3. Daubert, T. E., Hutchison, G., AIChE Symposium Series, 86 (279), 93 (1990).

FIGURE REFERENCES

3   Thermodynamics Research Center, "Selected Values of Properties of Chemical Compounds," Data Project, Texas A&M University, College Station, Texas (loose-leaf data sheets, extant, 1980).

39   Riddick J. A., Bunger, W. B., "Organic Solvents: Physical properties and Methods of Purification," 3rd ed., Wiley Interscience, New York (1970).

380   Lydersen, A. L., "Estimation of Critical Properties of Organic Compounds," University of Wisconsin Coll. Eng. Exp. Stn. Rept. 3, Madison, Wis. (April 1955).

1222   Toczylkin, L. S., Young, C. L., "Gas-Liquid Critical Temperatures of Mixtures Containing Electron Donors. II. Amine Mixtures," J. Chem. Thermo. 12, 365 (1980).

1391   Fuge, E. T. J., Bowden, S. T., Jones, W. J., "Some Physical Properties of Diacetone Alcohol, Mesityl Oxide and Methyl Isobutyl Ketone," J. Phys. Chem. 56, 1013 (1952).

1946   Majer, V., Svoboda, V., Koubek, J., Pick, J., "Temperature Dependence of Heats of Vaporization, Saturated Vapor Pressures, Cohesive Energies for a Group of Amines," Col. Czech. Chem. Comm. 44, 3521 (1979).

2668   Aldrich, "Handbook of Fine Chemicals," (Catalog) Aldrich Chemical Co., Milwaukee, WI. (1984-1985).

2867   Handbook of Data on Organic Compounds, Edited by R. C. Weast, M. J. Astle, CRC Press, Inc., Boca Raton, FL (1985).

3092   Ljunggren, G., "Die Hydrolyse-Geschwindigkeit des Acetessigs Auremethyl-Esters," Chem. Ber. 56, 2741 (1923).

3111   Fiest, A., "Ein Kondensations Produkt Aus Vier Acetessigeist Ermolekulen, Alpha (1,3 Dimethyl-2-Carboxyphenyl)-Beta-Methylglytaconsaure," Liebigs Ann. Chem. 433, 62 (1923).

4284   Adams, R., Rindfusz, R. E., "Cyclic Ethers from o-Allylphenols; Methyl Coumaranes," J. Amer. Chem. Soc. 41, 655 (1919).

4286   Gaertner, R., "The Equilibrium Between the o-Allephenoxide and 2-Benzoturylmethyl Anions," J. Amer. Chem. Soc. 73, 4400 (1950).

4340   Maitte, P., "Chroman and Isochroman, Synthesis of Chromenes," Ann. (Paris) 9, 431 (1954).

997X   Low pressure data

998X   High pressure data

9990   Regression Line

Table 1
Experimental Vapor Pressure Data

	Compound	T_{range} (°C)	P_{range} (kPa)	Notes
1.	sec-Butyl acetate	31.0 - 250.5	3.09 - 1983.2	
2.	t-Butyl acetate	28.2 - 245.7	6.25 - 5215.2	
3.	Methylacetoacetate	80.0 - 228.8	3.94 - 592.2	a
4.	Ethyl-t-butyl ether	20.7 - 211.6	9.81 - 2643.8	ab
5.	2-Phenylpropionaldehyde	102.2 - 237.9	3.81 - 235.1	d
6.	2-Methylbenzofuran	93.5 - 295.0	3.56 - 758.0	
7.	Diacetone alcohol	73.5 - 195.6	3.55 - 1458.6	c
8.	Ethylthioethanol	101.8 - 303.2	6.36 - 2313.6	
9.	n-Butylamine	19.3 - 241.3	9.27 - 3691.4	b
10.	Benzylamine	89.9 - 266.4	4.03 - 409.4	a
11.	N-Aminoethyl ethanolamine	174.3 - 260.4	12.4 - 244.1	
12.	Methane sulfonyl chloride	68.8 - 159.8	3.68 - 94.4	e
13.	1,2-Ethanedithiol	55.3 - 135.8	3.77 - 78.5	f

Notes

a - Sample started to decompose at upper temperature limit.
b - Could not go to lower pressure because of sample's low NBP.
c - Sample decomposing above 145°C.
d - Sample decomposed above upper temperature limit.
e - Sample reacted with mercury in capillary apparatus.
f - Sample unavailable for capillary apparatus.

Table 2
Experimental Data

SEC-BUTYL ACETATE

T(C)	PRESS.(PA)	REF.
31.000	3.086413D+03	9971
39.700	6.079500D+03	9971
43.400	7.546046D+03	9971
47.700	9.305901D+03	9971
53.700	1.224566D+04	9971
59.500	1.579203D+04	9971
65.600	2.052498D+04	9971
72.100	2.651115D+04	9971
78.800	3.419052D+04	9971
84.400	4.182323D+04	9971
87.700	4.976257D+04	9971
93.600	5.760193D+04	9971
98.100	6.637454D+04	9971
103.200	7.857354D+04	9971
111.200	1.170323D+05	9982
120.300	1.548035D+05	9982
130.500	1.998110D+05	9982
140.900	2.462328D+05	9982
151.000	3.153709D+05	9982
156.800	3.477503D+05	9981
161.700	3.929728D+05	9981
168.100	4.437629D+05	9981
176.800	5.386175D+05	9981
185.200	6.409434D+05	9981
192.000	7.251903D+05	9981
199.200	8.396703D+05	9981
209.500	1.003300D+06	9981
217.500	1.147728D+06	9981
227.400	1.352518D+06	9981
238.300	1.616594D+06	9981
250.500	1.983205D+06	9981

T-BUTYLACETATE

T(C)	PRESS.(PA)	REF.
28.200	6.252319D+03	9971
32.300	7.899350D+03	9971
35.900	9.552548D+03	9971
41.100	1.225233D+04	9971
47.100	1.579870D+04	9971
53.300	2.087828D+04	9971
59.100	2.631784D+04	9971
65.900	3.429051D+04	9971
71.200	4.198988D+04	9971
76.200	5.036252D+04	9971
80.100	5.754860D+04	9971
84.700	6.714781D+04	9971
89.200	7.792026D+04	9971
92.400	8.620620D+04	9971
95.300	9.411226D+04	9971
103.500	1.204103D+05	9981
111.400	1.535735D+05	9981
121.800	2.136975D+05	9981
130.400	2.902958D+05	9981
142.100	3.766576D+05	9981
151.900	4.804733D+05	9981
165.000	8.549848D+05	9982
182.200	1.299159D+06	9982
196.000	1.675563D+06	9982
211.800	2.616309D+06	9983
222.100	3.415907D+06	9983
233.900	4.358203D+06	9984
245.700	5.215233D+06	9984
115.900	2.334546D+05	9985
126.500	3.206797D+05	9985
135.500	3.971066D+05	9985
95.700	1.058391D+05	9986
100.400	1.114208D+05	9986
108.700	1.493299D+05	9986

METHYLACETOACETATE

T(C)	PRESS.(PA)	REF.
80.000	3.939676D+03	9971
89.000	6.139495D+03	9971
93.800	7.706033D+03	9971
99.400	9.725867D+03	9971
105.800	1.264563D+04	9971
111.800	1.601868D+04	9971
118.800	2.087162D+04	9971
125.600	2.669780D+04	9971
133.100	3.425718D+04	9971
139.300	4.193655D+04	9971
144.700	4.978924D+04	9971
148.300	5.797523D+04	9971
155.100	7.074751D+04	9971
160.600	8.322649D+04	9971
165.600	9.469221D+04	9971
185.500	2.121946D+05	9981
196.400	2.989531D+05	9981
206.500	3.974965D+05	9981
228.800	5.921537D+05	9981*
172.800	1.409660D+05	9982
179.700	1.859253D+05	9982
218.900	5.846965D+05	9982
165.800	1.132442D+05	9983
171.800	1.561785D+05	9983
179.900	2.054109D+05	9983

* SAMPLE DECOMPOSING AT AND ABOVE THIS POINT

ETHYL-T-BUTYL ETHER

T(C)	PRESS.(PA)	REF.
20.700	9.012520D+03	9971
24.500	1.243231D+04	9971
30.400	1.651198D+04	9971
34.500	2.086495D+04	9971
42.000	2.698445D+04	9971
44.700	3.477714D+04	9971
47.400	4.236318D+04	9971
52.200	5.054251D+04	9971
56.200	5.839520D+04	9971
60.200	6.723447D+04	9971
64.600	7.841355D+04	9971
67.700	8.692618D+04	9971
70.900	9.576546D+04	9971
161.000	8.760055D+05	9981
172.500	1.084594D+06	9981
180.500	1.261445D+06	9981
190.500	1.502935D+06	9981
201.200	2.029094D+06	9981*
211.600	2.643845D+06	9981*
72.600	1.020933D+05	9982
82.000	1.290978D+05	9982
90.500	1.754241D+05	9982
100.100	2.389686D+05	9982
110.900	3.019583D+05	9982
120.200	3.687674D+05	9982
131.000	4.947271D+05	9982
141.600	6.290578D+05	9982
151.600	7.721763D+05	9982
70.000	8.530564D+04	9983
73.900	1.045166D+05	9983
77.300	1.161382D+05	9983
87.200	1.618972D+05	9983

* SAMPLE DECOMPOSING AT AND ABOVE THIS POINT

Dl-2-PHENYLPROPIONALDEHYDE

T(C)	PRESS.(PA)	REF.
102.200	3.806354D+03	9971
109.800	6.392808D+03	9971
113.100	7.992676D+03	9971
120.800	1.143906D+04	9971
130.100	1.553872D+04	9971
138.400	2.058497D+04	9971
147.500	2.643783D+04	9971
157.000	3.373723D+04	9971
166.800	4.208321D+04	9971
172.500	4.954259D+04	9971
176.800	5.798190D+04	9971
182.400	6.652120D+04	9971
187.300	7.843355D+04	9971
190.800	1.034686D+05	9982
193.900	6.992294D+04	9981
199.900	1.188861D+05	9982
203.100	1.110241D+05	9981
212.400	1.348505D+05	9981
224.800	1.670455D+05	9981
237.900	2.351036D+05	9981

2-METHYLBENZOFURAN

T(C)	PRESS.(PA)	REF.
93.500	3.559707D+03	9971
107.100	6.119497D+03	9971
112.800	7.686035D+03	9971
118.400	9.932516D+03	9971
124.500	1.193902D+04	9971
131.400	1.585203D+04	9971
139.400	2.073829D+04	9971
146.900	2.650449D+04	9971
155.200	3.412386D+04	9971
162.100	4.202988D+04	9971
167.800	4.993589D+04	9971
173.400	5.749527D+04	9971
178.600	6.600790D+04	9971
184.600	7.778694D+04	9971
184.600	8.360416D+04	9982
195.000	9.448294D+04	9982
206.300	1.232585D+05	9982
207.800	1.452927D+05	9981
217.500	1.880083D+05	9981
228.200	2.358500D+05	9981
239.000	2.953450D+05	9981
250.700	3.590994D+05	9981
262.700	4.241388D+05	9981
276.700	5.615758D+05	9981
295.000	7.579677D+05	9981

DIACETONE ALCOHOL

T(C)	PRESS.(PA)	REF.
73.500	3.546375D+03	9972
84.600	6.026171D+03	9972
89.700	7.639372D+03	9972
93.700	9.492553D+03	9972
100.300	1.258563D+04	9972
106.900	1.621200D+04	9972
119.200	2.649115D+04	9972
125.900	3.404387D+04	9972
131.100	4.183656D+04	9972
135.200	5.003588D+04	9972
138.300	5.782191D+04	9972
142.700	6.688783D+04	9972
145.700	7.836022D+04	9972*
147.100	8.661288D+04	9972*
148.800	9.390561D+04	9972*
156.600	3.042453D+05	9989*
158.700	2.593662D+05	9986*
165.500	3.598827D+05	9989*
171.100	4.015641D+05	9986*
175.400	5.025996D+05	9989*
181.300	6.265936D+05	9986*
185.900	9.214039D+05	9989*
189.500	8.221232D+05	9986*
195.600	1.458644D+06	9989*

* SAMPLE DECOMPOSING AT AND ABOVE THIS POINT

ETHYLTHIOETHANOL

T(C)	PRESS.(PA)	REF.
101.800	6.359477D+03	9971
107.400	8.072669D+03	9971
112.100	9.812526D+03	9971
118.100	1.279228D+04	9971
124.700	1.654531D+04	9971
131.300	2.116493D+04	9971
138.400	2.707111D+04	9971
145.600	3.473714D+04	9971
151.800	4.226319D+04	9971
157.100	5.033586D+04	9971
161.900	5.826187D+04	9971
166.400	6.693450D+04	9971
171.800	7.830689D+04	9971
175.700	8.723949D+04	9971
178.900	9.505218D+04	9971
183.500	1.242858D+05	9981
193.500	2.058620D+05	9981
203.500	2.519722D+05	9981
212.800	3.391697D+05	9981
223.300	4.256378D+05	9981
233.100	5.519615D+05	9981
242.800	6.535962D+05	9981
253.700	7.424014D+05	9981
263.000	9.958422D+05	9982
272.800	1.195304D+06	9982
282.900	1.609891D+06	9982
292.500	1.831898D+06	9982
303.200	2.313572D+06	9982
179.900	1.146430D+05	9983
182.500	1.296304D+05	9983
190.900	1.707263D+05	9983
200.500	2.041361D+05	9983

BUTYLAMINE

T(C)	PRESS.(PA)	REF.
19.300	9.272571D+03	9971
24.800	1.246564D+04	9971
30.100	1.626533D+04	9971
35.400	2.091828D+04	9971
40.800	2.666447D+04	9971
46.800	3.444383D+04	9971
51.600	4.619620D+04	9971
56.100	5.019587D+04	9971
59.900	5.788191D+04	9971
63.800	6.705449D+04	9971
68.200	7.850688D+04	9971
71.200	8.676620D+04	9971
73.700	9.429225D+04	9971
82.000	1.057828D+05	9981
97.600	2.074750D+05	9981
107.200	2.674804D+05	9981
117.700	3.881050D+05	9981
128.600	4.660975D+05	9981
139.200	6.192121D+05	9981
148.300	6.927063D+05	9981
90.700	1.338911D+05	9982
110.900	2.890580D+05	9982
157.700	8.167198D+05	9983

T(C)	PRESS.(PA)	REF.
168.200	1.064396D+06	9983
179.000	1.377591D+06	9983
189.800	1.697666D+06	9983
200.800	1.866104D+06	9984
211.100	2.141460D+06	9984
221.800	2.780218D+06	9984
231.800	3.076338D+06	9984
241.300	3.691402D+06	9984
74.800	9.993785D+04	9985
79.500	1.172441D+05	9985
101.500	2.168830D+05	9985
121.400	4.257247D+05	9985
131.000	5.081071D+05	9985
141.100	6.449312D+05	9985
151.200	8.217033D+05	9985
84.400	1.183078D+05	9986
94.500	1.641370D+05	9986

BENZYL AMINE

T(C)	PRESS.(PA)	REF.
89.900	4.026336D+03	9971
99.300	6.052836D+03	9971
106.100	8.225990D+03	9971
111.000	1.001918D+04	9971
116.100	1.229232D+04	9971
123.500	1.639199D+04	9971
130.000	2.057831D+04	9971
137.300	2.653115D+04	9971
145.800	3.474381D+04	9971
152.000	4.235652D+04	9971
157.800	5.021587D+04	9971
162.700	5.786857D+04	9971
167.900	6.596791D+04	9971
173.700	7.784026D+04	9971
178.100	8.717283D+04	9971
181.300	9.451223D+04	9971
202.300	1.259118D+05	9981
212.400	1.760573D+05	9981
222.200	2.366930D+05	9981
232.700	2.801702D+05	9981
245.700	3.831774D+05	9981
256.200	4.271051D+05	9981
266.400	4.093881D+05	9982*
189.100	1.004193D+05	9983
196.300	1.216064D+05	9983
217.700	2.071608D+05	9983

* SAMPLE DECOMPOSING AT AND ABOVE THIS POINT

N-AMINOETHYL ETHANOLAMINE

T(C)	PRESS.(PA)	REF.
174.300	1.236565D+04	9971
180.500	1.582537D+04	9971
187.700	2.094494D+04	9971
195.500	2.661114D+04	9971
202.400	3.433718D+04	9971
210.200	4.203654D+04	9971
215.100	5.029586D+04	9971
218.200	5.764859D+04	9971
225.700	6.646787D+04	9971
230.900	7.819357D+04	9971
230.500	8.273287D+04	9982
234.600	1.114743D+05	9981
240.500	1.434156D+05	9982
242.200	1.331901D+05	9981
250.500	1.941013D+05	9982
250.700	1.596024D+05	9981
260.400	2.441495D+05	9982

METHANESULFONYL CHLORIDE

T(C)	PRESS.(PA)	REF.
68.800	3.679697D+03	9971
80.800	6.366143D+03	9971
85.400	7.859354D+03	9971
90.700	9.772530D+03	9971
97.100	1.262563D+04	9971
103.700	1.625200D+04	9971
111.200	2.139824D+04	9971
117.500	2.678446D+04	9971
125.600	3.523710D+04	9971
130.900	4.197655D+04	9971
136.600	4.996256D+04	9971
141.200	5.766192D+04	9971
145.900	6.653453D+04	9971
152.200	7.792692D+04	9972
152.900	8.123332D+04	9971
153.300	8.127332D+04	9972
154.700	8.459971D+04	9972
157.300	8.855938D+04	9972
158.600	9.155247D+04	9972
159.800	9.439890D+04	9972

1,2-ETHANEDITHIOL

T(C)	PRESS.(PA)	REF.
55.300	3.766357D+03	9971
66.700	6.399474D+03	9971
71.100	7.926015D+03	9971
75.700	9.572546D+03	9971
81.900	1.239898D+04	9971
89.100	1.657197D+04	9971
95.200	2.106493D+04	9971
101.900	2.673113D+04	9971
109.100	3.415052D+04	9971
115.500	4.238985D+04	9971
120.600	4.980257D+04	9971
125.600	5.892521D+04	9971
130.300	6.714114D+04	9971
135.800	7.853354D+04	9971

Table 3
Coefficients for Regression Equation

		A	B	C	D	E†	Average % Error
1. sec-Butyl acetate	A	75.323951	-7720.885919	-7.574419	1.535706x10⁻¹⁷	6	7.5
	B	48.512543	-5991.582962	-3.588045	-6.222879x10⁻¹⁸	6	3.9
2. t-Butyl acetate	A	115.8667444	-9169.3898175	-13.6937258	1.06377x10⁻⁵	2	13.0
	B	-28.3442983	-1932.0846476	7.6330158	-9.283322x10⁻¹⁸	6	5.3
3. Methylacetoacetate	A	105.4280302	-9545.1031257	-12.1001775	6.77959x10⁻⁶	2	18.1
	B	58.1490747	-8099.6275953	-4.6001637	-8.958832x10⁻¹⁸	6	9.6
4. Ethyl-t-butyl ether	A	73.6502157	-6498.2215970	-7.4534566	2.580403x10⁻¹⁷	6	13.8
	B	127.4739176	-9226.9763593	-15.2861504	5.216041x10⁻¹⁷	6	4.5
5. 2-Phenylpropionaldehyde	A	102.433358	-9698.33245	-11.629701	5.8179x10⁻⁶	2	9.2
	B	186.032939	-15066.354388	-23.202541	2.516307x10⁻¹⁷	6	7.4
6. 2-Methylbenzofuran	A	98.209725	-9444.217115	-11.012118	5.21872x10⁻⁶	2	5.8
	B	49.7806	-7257.841181	-3.696873	-5.646065x10⁻¹⁹	6	3.2
7. Diacetone alcohol**	A	112.25305	-9769.082314	-13.120133	8.2000x10⁻⁶	2	28.2
	B	-402.065219	14927.08452	62.628104	0	6	19.3
8. Ethylthioethanol	A	116.0993819	-10605.7649509	-13.5199303	7.65857x10⁻⁶	2	11.9
	B	-90.4797514	216.7536127	16.6628272	-3.276045x10⁻¹⁷	6	5.3
9. n-Butylamine	A	92.8417028	-6970.7866518	-10.6659943	8.70734x10⁻⁶	2	6.3
	B	58.2430884	-5845.0986324	-5.1556808	1.17694x10⁻⁶	2	4.1
10. Benzylamine	A	100.2150190	-9479.1002852	-11.2869927	5.5668x10⁻⁶	2	6.5
	B	97.0647042	-9628.6076612	-10.5529112	1.1449x10⁻¹⁷	6	4.9
11. N-Aminoethyl ethanolamine	A	100.821905	-12390.626199	-10.445103	0	6	7.9
	B	*	*	*	*	*	*
12. Methanesulfonylchloride	A	96.0801	-8931.130334	-10.706579	5.3600x10⁻⁶	2	4.1
	B	96.136012	-9226.829486	-10.449221	1.192325x10⁻¹⁷	6	0.8
13. 1,2-Ethanedithiol	A	90.799699	-8170.5402626	-10.0518267	5.3311x10⁻⁶	2	0.9
	B	80.6698534	-8052.2689402	-8.2725861	8.6454x10⁻¹⁸	6	0.3

A - Constrained run
B - Unconstrained run
† - E chosen as 2 or 6 according to which gave better data fit
* - Program would not run unconstrained for this compound [singular matrix]
** - Regression only of low pressure data - error analysis of all data

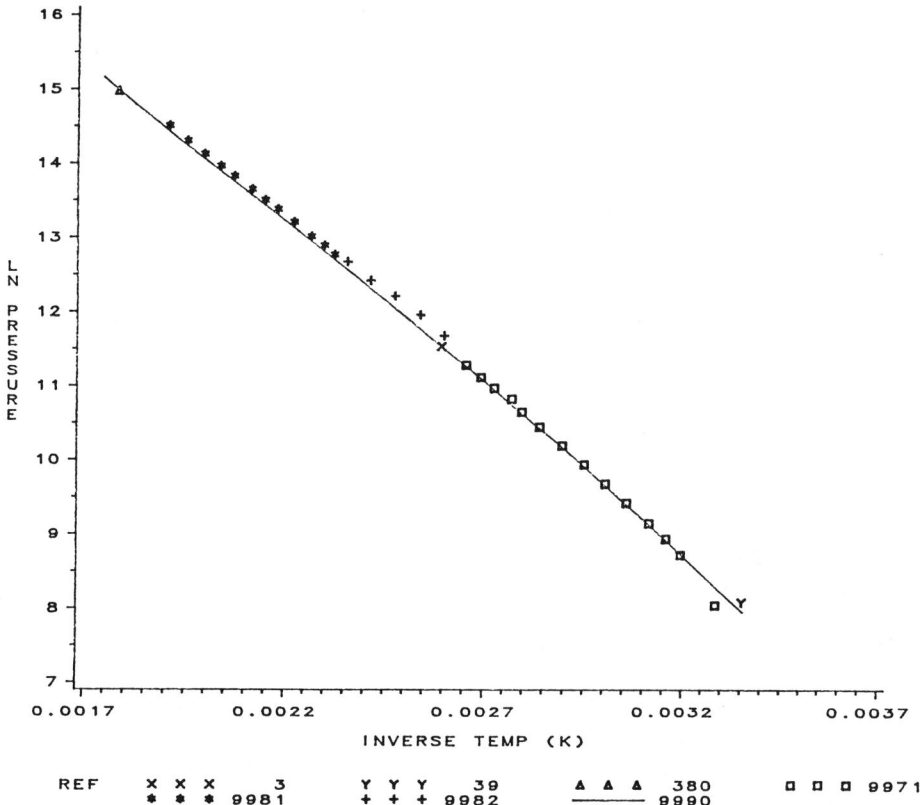

Figure 1. Experimental and literature data with constrained regression line for sec-butyl acetate

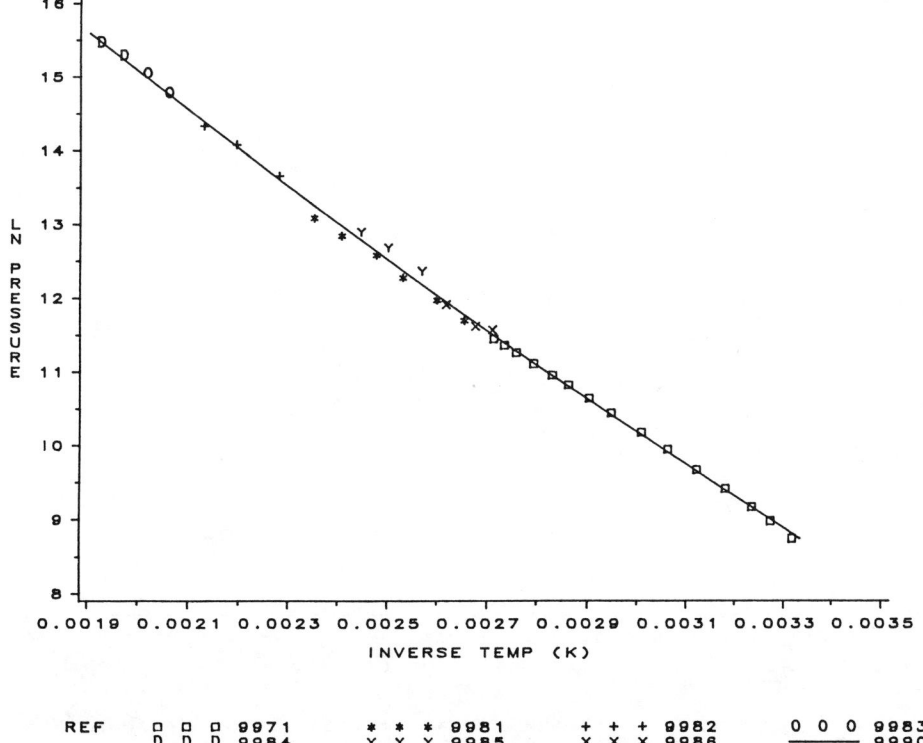

Figure 2. Experimental data with unconstrained regression line for *t*-butylacetate

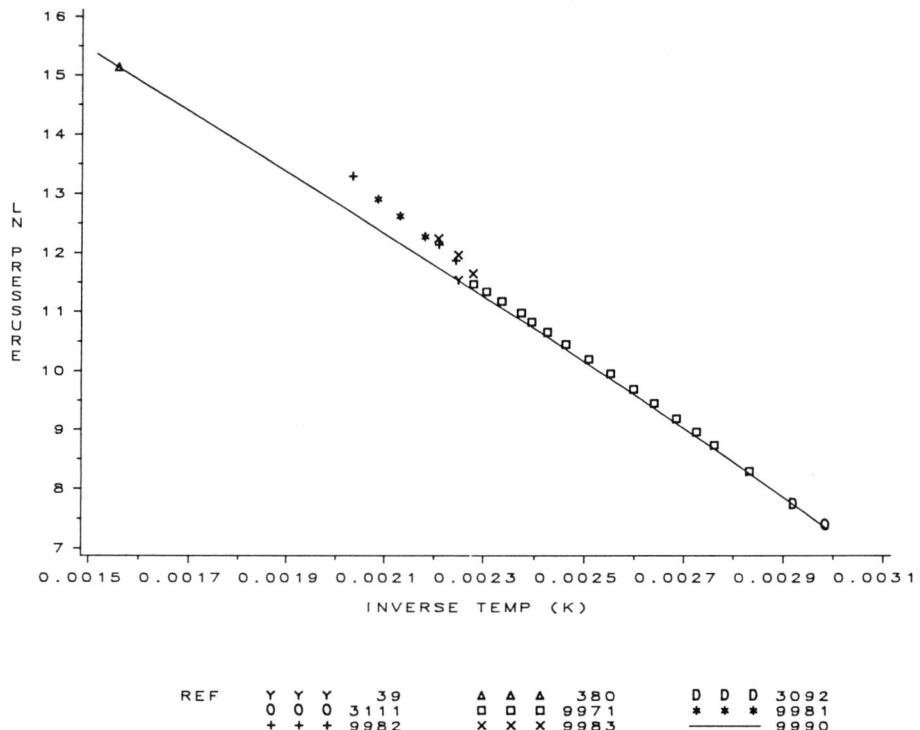

Figure 3. Experimental and literature data with constrained regression line for methylacetoacetate

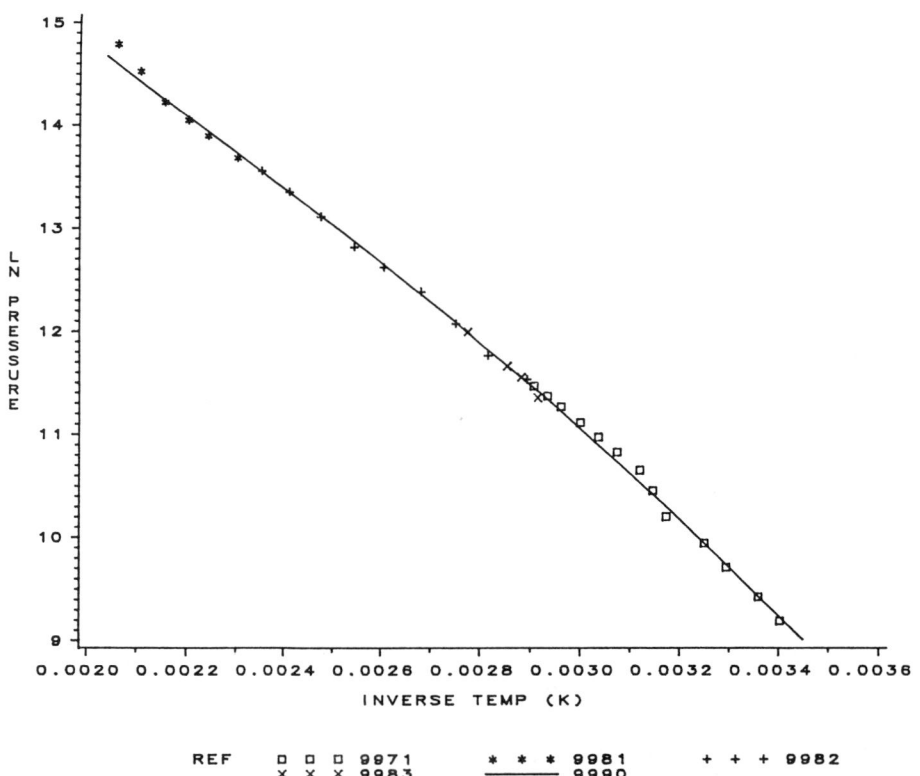

Figure 4. Experimental data with unconstrained regression line for *t*-butyl ethyl ether

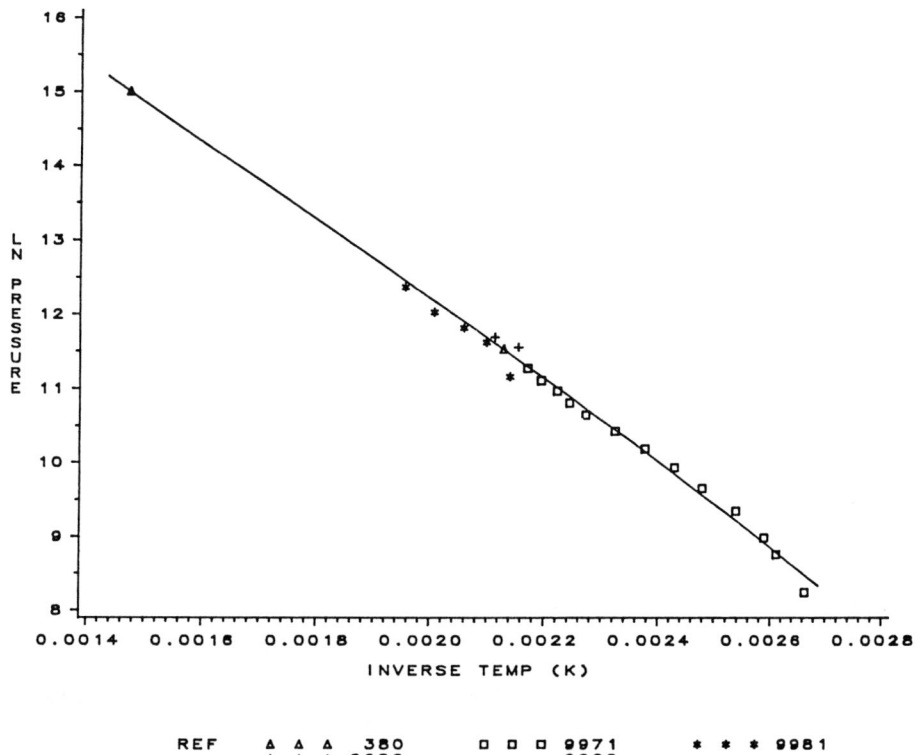

Figure 5. Experimental data with constrained regression line for 2-phenylpropionaldehyde

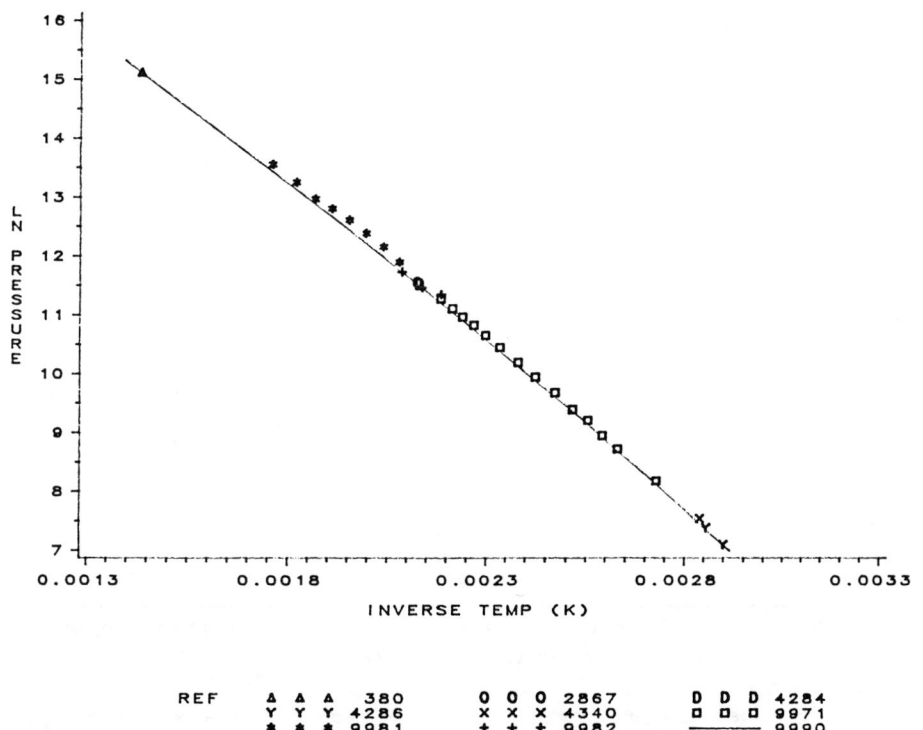

Figure 6. Experimental and literature data with constrained regression line for 2-methylbenzofuran

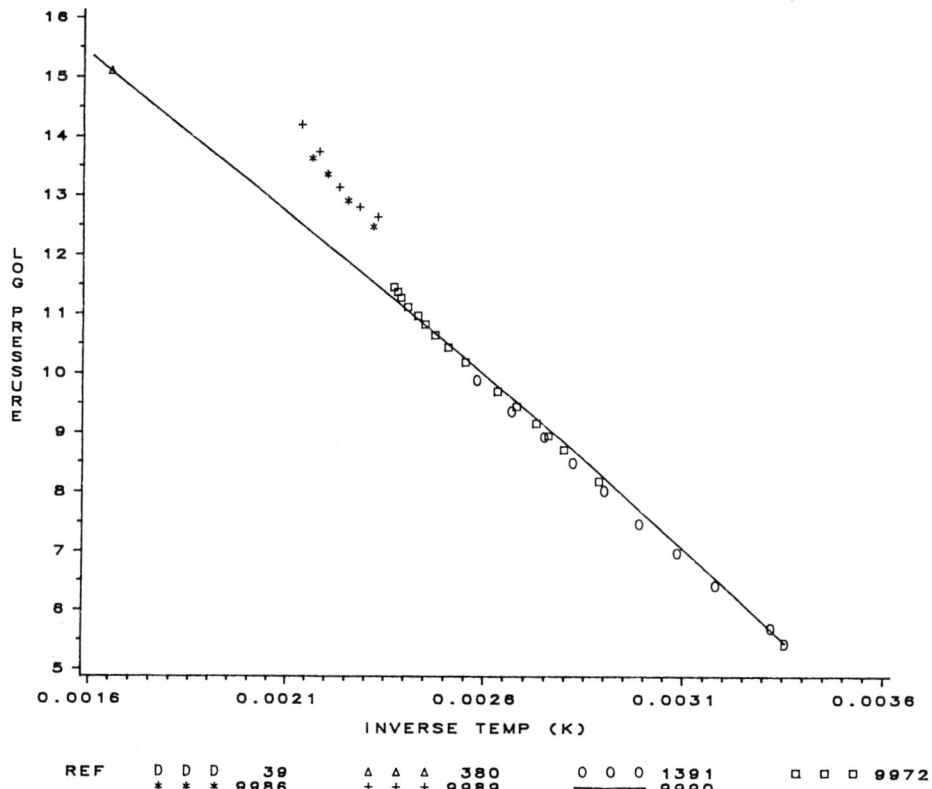

Figure 7. Experimental and literature data with constrained regression line for diacetone alcohol

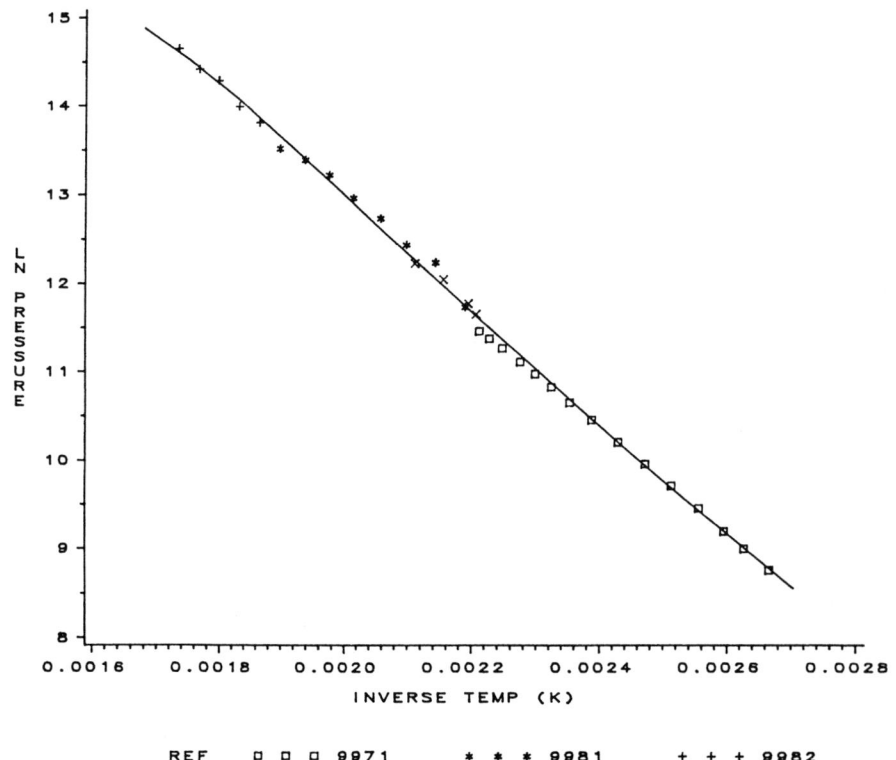

Figure 8. Experimental data with unconstrained regression line for ethyl thioethanol

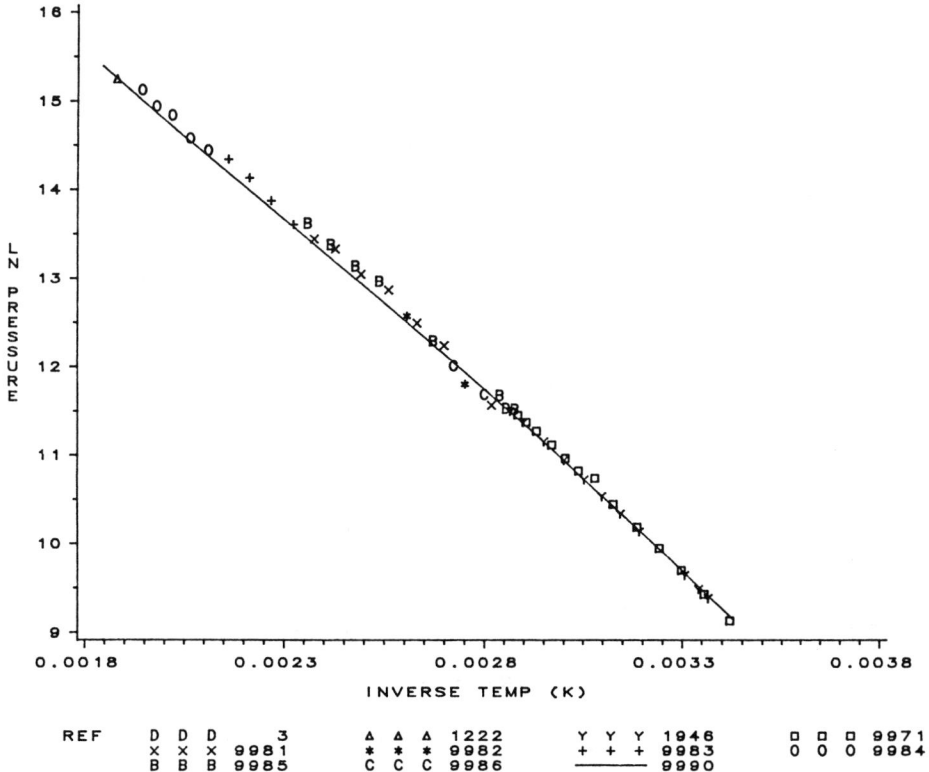

Figure 9. Experimental and literature data with constrained regression line for *n*-butylamine

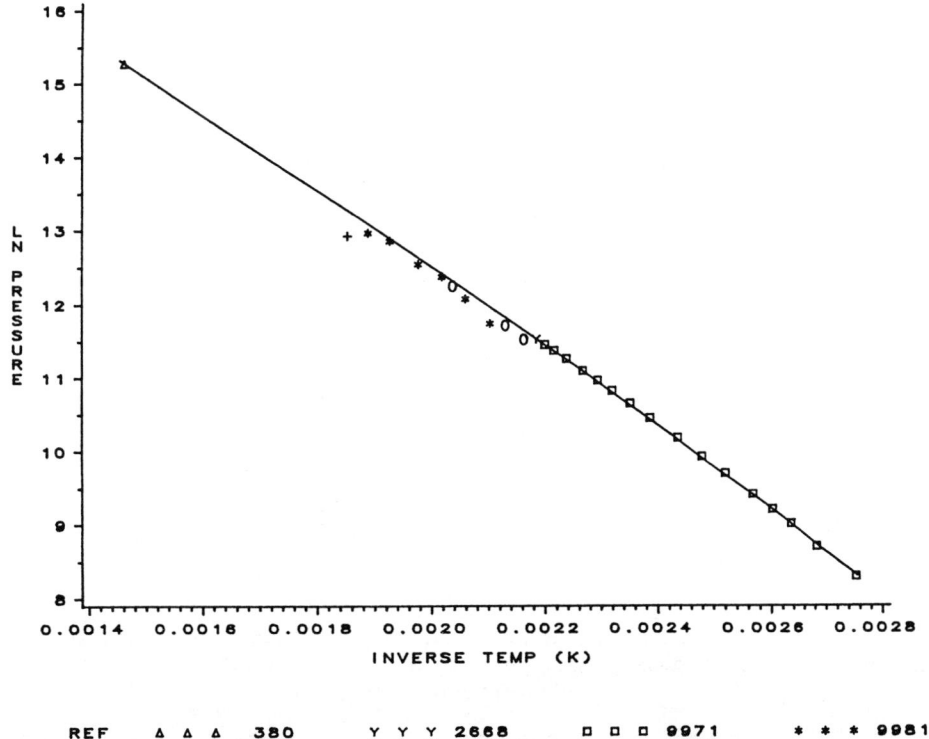

Figure 10. Experimental and literature data with constrained regression line for benzylamine

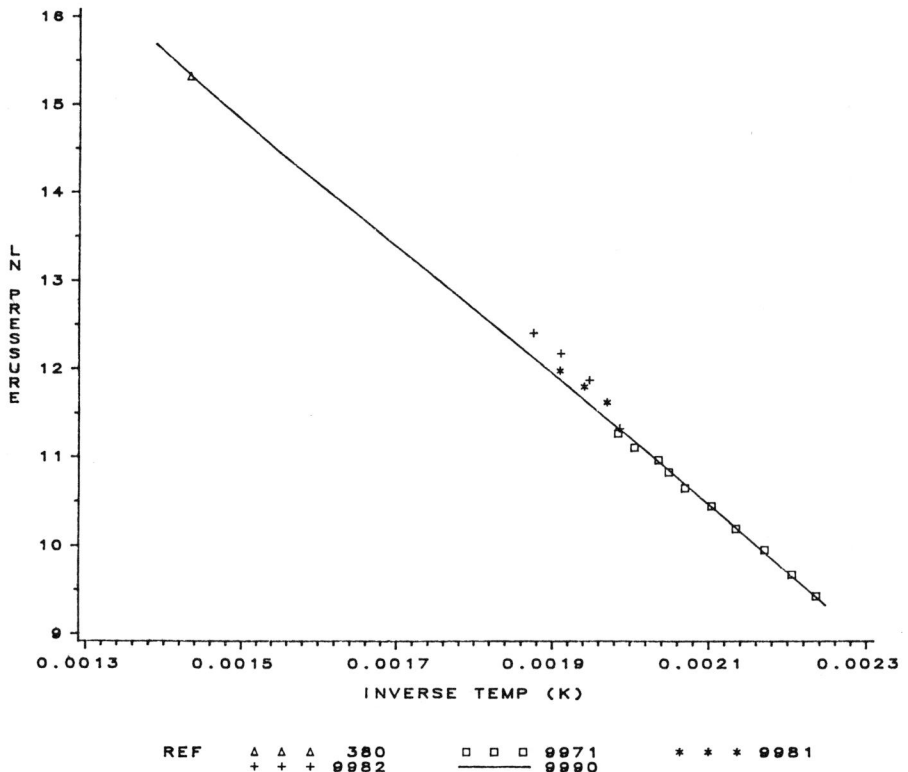

Figure 11. Experimental data with constrained regression line for *N*-aminoethyl ethanolamine

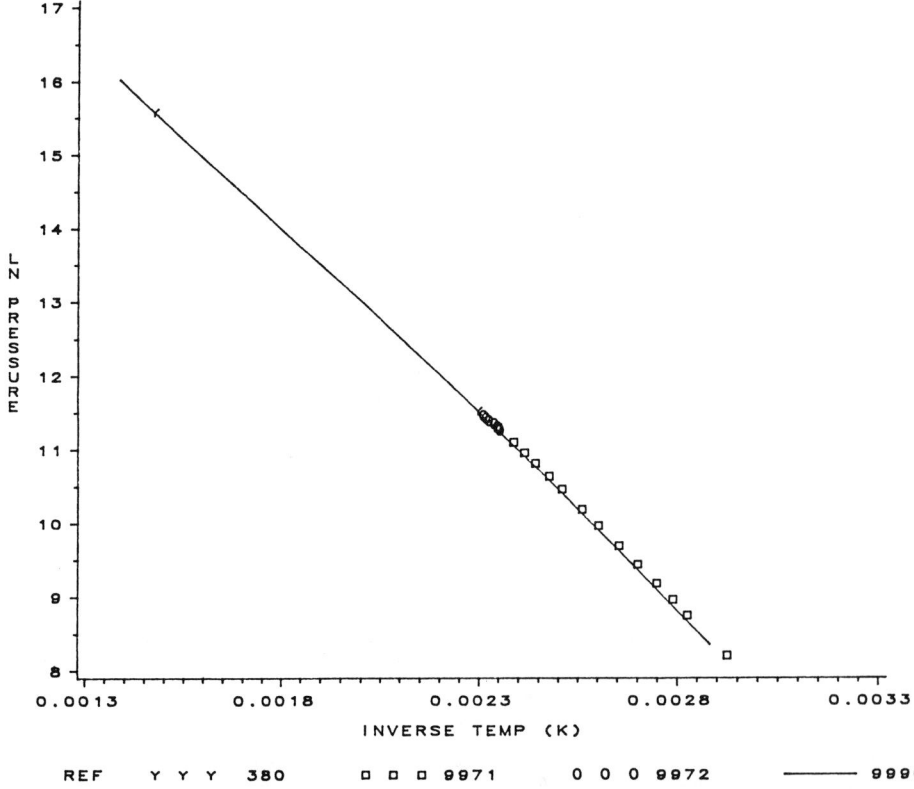

Figure 12. Experimental data with constrained regression line for methanesulfonyl chloride

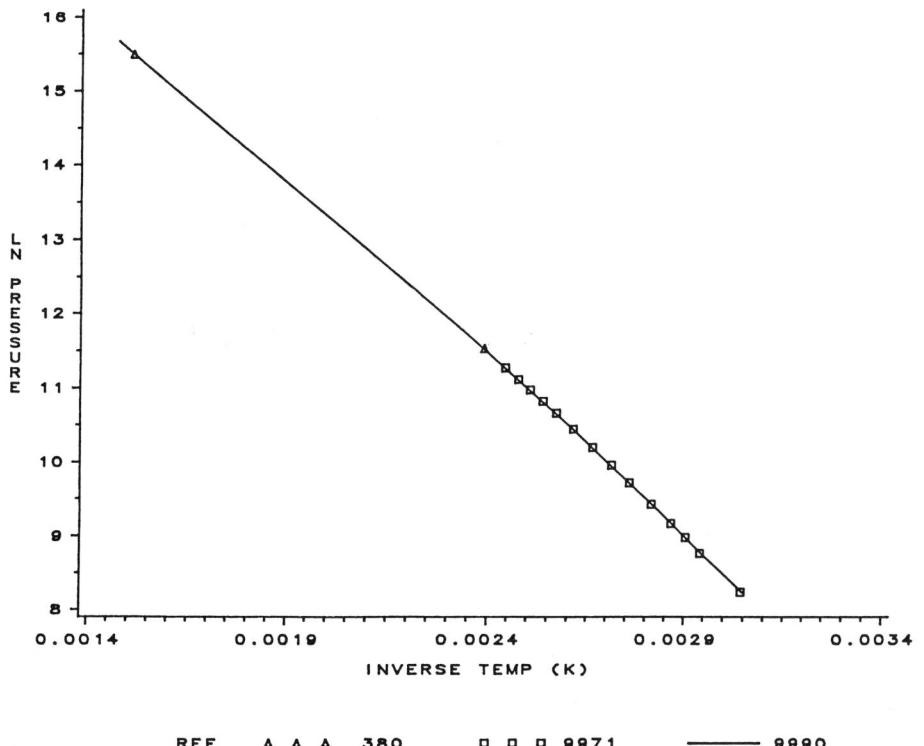

Figure 13. Experimental data with constrained regression line for 1,2-ethanedithiol

THE CRITICAL PRESSURES AND TEMPERATURES OF TEN SUBSTANCES USING A LOW RESIDENCE TIME FLOW APPARATUS

Amyn S. Teja and Daniel J. Rosenthal ■ School of Chemical Engineering,
Georgia Institute of Technology,
Atlanta, Georgia 30332-0100

The critical pressures and temperatures of acetophenone, 2-propoxyethanol, 2-butoxyethanol, ethylene glycol, ethyl-3-ethoxy propionate, 1-methoxy 2-propanol acetate, 2-(2-propoxyethoxy) ethanol, 2-(2-butoxyethoxy) ethanol, monoethanolamine and valeric acid were measured using a low residence time flow technique developed in our laboratory. The method minimizes the effect of thermal decomposition because of the short time spent at elevated temperatures and yields reliable and accurate values of the critical properties in many cases.

In spite of many studies of the critical properties of pure fluids reported in the literature (see, for example, the compilation of Ambrose [1]), reliable experimental data are still lacking for many substances. This is particularly true for materials that decompose or associate before attaining their critical states.

As part of a continuing effort in the area of critical point measurement in our laboratory, we have previously reported the critical temperatures and densities of a large number of substances of industrial interest [2-8]. We have also described a low residence time flow technique for the measurement of the critical temperature and pressures of thermally stable and unstable fluids [8, 9].

New results for the critical properties of ten substances using the flow apparatus are presented in this paper and comparisons between the two methods developed in our laboratory are shown. The ten substances were chosen from a list of chemicals of interest to the industrial sponsors of this project and because their estimated critical properties were within the range of capability of the equipment.

Dan Rosenthal is now with the Solway Corporation in Brussels, Belgium.

EXPERIMENTAL

In the flow method, the fluid of interest is pumped rapidly through a heated view cell at a temperature and pressure which result in critical opalescence and/or meniscus disappearance or reappearance in the observation chamber. The residence time of the fluid at elevated temperatures is kept low in order to minimize thermal decomposition or association.

A schematic diagram of the flow apparatus is shown in Fig. 1. An Aldex high pressure pump was used to pump a degassed sample of the fluid under study through a heated view cell, where the pressure and temperature of the fluid were measured and any phase changes noted. The view cell was made of stainless steel and had two windows made of borosilicate glass to admit light into the chamber and for visual observation of any phase changes. The internal volume of the cell was 4.36 ml. Two type K thermocouples were inserted into the observation chamber at distances of approximately one-third of the length of the cell from each end.

The cell was placed inside a thermostatted air bath where its temperature could be maintained constant within 0.1 K. The pressure in the cell, which was measured using a calibrated digital Heise gauge, could be adjusted using a micro-metering valve located downstream from

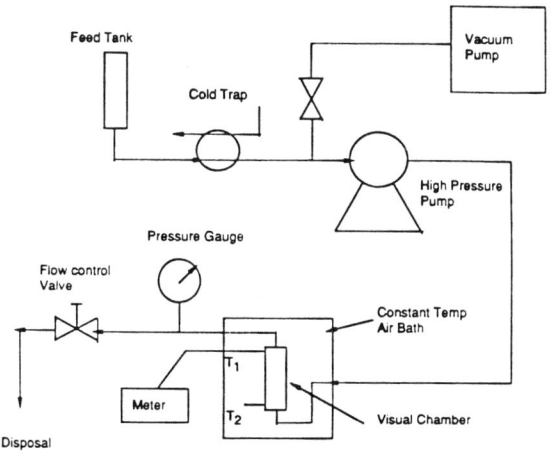

Fig. 1 Schematic diagram of the apparatus.

the cell. The fluid leaving the cell through the micro-metering valve was discarded.

At the beginning of the experiment, the air bath temperature was set at a value approximately 20 K above the estimated critical temperature of the fluid and the fluid was pumped into the system with the micro-metering valve closed completely in order to build up pressure in the system. When the system pressure approached the critical pressure of the fluid, a flat meniscus separating the vapor and liquid phases could be observed in the observation chamber, provided the temperature inside the chamber was close to the critical temperature of the fluid. If the temperature inside the chamber was much lower than the critical temperature, then the meniscus had pronounced curvature because of the difference in density between the vapor and liquid phases. It should be added here that the temperature of the fluid in the cell was less than the preset temperature of the air bath. When the pressure exceeded the critical pressure, the micro-metering valve was adjusted in order to obtain steady flow and a decrease of the system pressure. Pressure reduction in the critical region led to the appearance of an orange-red band, which was first observed in the fluid at pressures slightly above the critical. The pressure was recorded when the color was the most intense (maximum opalescence) or when meniscus disappearance could be observed in the cell. The position of the red band in the observation chamber could be manipulated by changing the flow rate of the fluid. The flow rate was therefore varied until the red band moved to a position near one of the thermocouples. The temperature measured by that thermocouple was assumed to be the critical temperature of the fluid and was recorded. The typical temperature gradient in the cell was on the order of 5 K and the thickness of the band varied from a few mm for the stable substances to about 1 cm for the most unstable substances. Further details of the method and sources of errors are given elsewhere [9].

The accuracy of our measured critical temperatures is estimated to be ± 0.6 K for the stable substances and that of the pressure is estimated to be ± 0.02 MPa. The method is generally less accurate for thermally unstable substances. The precision of the temperature and pressure measurements was 0.1 K and 0.01 MPa, respectively.

In the case of unstable substances, decomposition or reaction cannot be prevented despite the low residence time in the heated zone. Figures 2 and 3 show the critical temperature and critical pressure vs. residence time behavior in this case. A change of critical temperature or pressure with residence time signifies thermal decomposition or association of the fluid and the data obtained must be extrapolated back to zero residence time to obtain the critical properties of the pure fluid as shown. An extrapolation

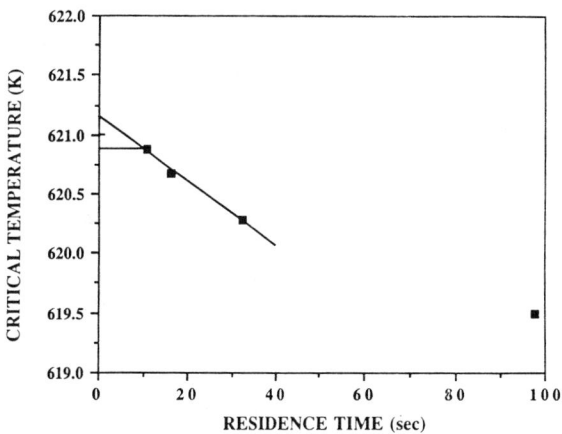

Fig. 2 Critical temperature vs residence time curve for ethyl-3-ethoxy propionate (EEP)

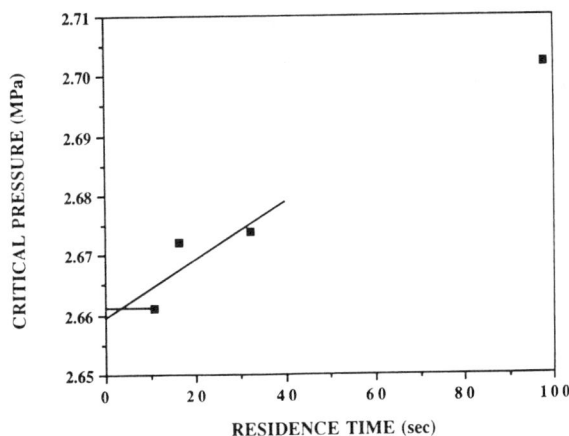

Fig. 3 Critical pressure vs residence time curve for ethyl-3-ethoxy propionate (EEP).

technique proposed in our previous work [4] was adopted in the present work. The actual critical temperature (or pressure) is assumed to be the average of the extrapolated temperature (or pressure) at zero residence time and the temperature (or pressure) at the first data point. The error is assumed to be the larger of 0.6 K (or 0.02 MPa) and half the difference between the two temperatures (or pressures). This extrapolation procedure assumes that the pure substance and its products of decomposition yield a linear critical locus, and was used with some success in our earlier studies. Note, however, that the extrapolation is carried out over a much smaller time interval in the flow studies compared with our previous sealed ampoule studies.

SOURCE AND PURITY OF THE SUBSTANCES STUDIED

The purest commercially available materials were used in the experiments and all chemicals were used without further purification. Table 1 gives the stated purity and supplier of the materials used. Each sample, however, was thoroughly degassed using a series of freeze-thaw - pump cycles before use.

RESULTS

Ten substances were studied in the present work. We have previously shown that reliable values of the critical properties can be obtained using the flow technique by comparing our measurements with experimental values reported in the literature. The results obtained in the present study are shown in Table 2 and detailed comments on each substance are given below.

Acetophenone exhibited a very slight decrease in critical temperature and increase in critical pressure with residence time in the flow apparatus. However, there were no visual signs of decomposition even at the longest residence times. We have previously found that acetophenone associated (T_c increased by 0.6 K in 20 minutes) in our sealed ampoule studies. This change would not be observable in the flow experiment because of the short residence times. Also, because of the high critical pressure and critical temperature of acetophenone, the flow-cell leaked and only a few data points could be obtained in the present study. Nevertheless, the value of the critical temperature obtained in the present study (709.6 K) is in excellent agreement with that obtained in the sealed ampoule study (709.5 K).

Both 2-butoxyethanol and 2-propoxyethanol were stable in the flow experiments, but exhibited mild decomposition in the static experiments. However, both methods give the same critical temperature, within experimental error Ethylene glycol changed color (indicative of a reaction) to orange in the flow experiments. However, no supercritical phase could be observed up to 9 MPa and 720 K. Because of the limitations of the apparatus, no further pressure increases were attempted. Nevertheless, it was possible to measure a vapor pressure point of 7.074 MPa at 694.9 K. The critical point lies above this measured point in P-T co-ordinates. (The critical temperature found in the sealed ampoule method was 718 ± 9 K and Lyons [10] reported a value of 13.1 MPa for the critical pressure.)

Our earlier studies showed that ethyl-3- ethoxy propionate (EEP) initially decomposes and then associates in the sealed ampoule experiments. The residence time in the flow apparatus is much smaller than in the sealed ampoule technique. Consequently, only the decomposition phase was observed for EEP in the flow experiments. Indeed, small increases in critical pressure and small decreases in critical temperature were observed with residence time. The extrapolated value obtained in the sealed ampoule method (618.7 K) is 2.3 K lower than the value obtained

(621.0 K) in the flow experiments. However, the value obtained in the sealed ampoule method depends strongly on the extrapolation procedure. A linear extrapolation using the second and third points of the temperature-time curve from the sealed ampoule experiments leads to a critical temperature of 620.6 ± 3 K. This value is in very good agreement with the critical temperature obtained in the present work (621.0 ± 0.6 K).

Because 2- (2- propoxyethoxy) ethanol and 2-(2-butoxyethoxy) ethanol decomposed rapidly, the critical point was obtained from the visual observation of the disappearance of the meniscus. 2-(2-propoxyethoxy) ethanol also decomposed rapidly in the sealed ampoule experiments and its critical temperature could only be estimated within ± 13 K in the static experiments. Much lower error bounds were obtained in the present study. 2- (2-butoxyethoxy) ethanol decomposed so rapidly at the highest flow rates that the pressure could not be controlled satisfactorily with the flow control valve. Nevertheless, both the critical temperature and the critical pressure could be obtained in the present work, although the error bounds were somewhat larger than ±0.02 MPa for the critical pressure.

Monoethanolamine decomposed rapidly in the flow experiment. However, its critical temperature could still be measured and was found to be 671.4 K. In the sealed ampoule experiment, decomposition was so extensive that the ampoule exploded. We previously estimated the critical temperature to be greater than 670 K. Lyons [10] reported a critical temperature of 678 ± 3.8 K and critical pressure of 7.12 MPa (by extrapolation of the vapor pressure curve). The critical pressure of monoethanolamine obtained in the present work is 8.03 MPa. Based on our sealed ampoule and flow studies, we believe that our values for the critical temperature and pressure of monoethanolamine are more reliable than the extrapolated values of Lyons.

Valeric acid was found to be highly unstable in the flow experiments. A thick cloud-like formation with an orange opalescent tint was observed near the critical point of the acid. Also, the fluid exiting the observation chamber was found to have a dark orange-red color. The effluent stream was therefore subjected to gas chromatographic analysis, but no decomposition products could be detected. It is possible that both the pure acid and the decomposition products had the same retention time in the capillary column, or that the concentration of the decomposition products was too small to be detected by gas chromatography.

Finally, 1-methoxy-2-propanol acetate was found to decompose only slightly at the higest flow-rates. Excellent values of the critical temperature and pressure could therefore be obtained in this work. Indeed, our critical temperature value for this substance agrees within experimental error with that obtained in the sealed ampoule experiments.

CONCLUSIONS

New data for the critical temperatures and pressures of ten substances have been obtained using the flow technique. The critical temperature data generally agree with those obtained in our sealed ampoule experiments. In the case of monoethanolamine, however, our values for both the critical temperature and critical pressure are significantly different from the literature values [10].

LITERATURE CITED

1. Ambrose D., "Vapour-liquid critical properties" National Physical Laboratory Report Chem. 107, 1980.

2. Smith R.L., M. J. Anselme, and A. S. Teja, *Fluid Phase Equilibria,* **31**, 161 (1986).

3. Smith R.L., W. B. Kay and A.S. Teja, *AIChEJ,* **33**, 232 (1987).

4. Anselme M. J., and A.S. Teja, *Fluid Phase Equilibria,* **40**, 127 (1988).

5. Teja A.S., and M. J. Anselme, *AIChE Symp. Ser.,* **86** (279), 115 (1990).

6. Teja, A.S., and M. J. Anselme, *AIChE Symp. Ser.,* **86** (279), 122 (1990).

7. Anselme, M. J. and A. S. Teja, *AIChE Symp. Ser.,* **86** (279), 128 (1990).

8. Teja, A. S. and D. J. Rosenthal, *AIChE Symp. Ser.,* **86** (279), 133 (1990).

9. Rosenthal D.J., and A.S. Teja, *AIChEJ,* **35**, 1829 (1989).

10. Lyons R.L., "The Determination of Critical Properties and Vapor Pressure of Thermally Stable and Unstable Compounds", M.S. Thesis, Pennsylvania State University, (1985).

Table 1 Source and purity of the substances studied.

Substance	Source	Purity, mol%
acetophenone	Aldrich	99.0
2-butoxyethanol	Kodak	99.5
2-propoxyethanol	Kodak	99.7
ethylene glycol	Fisher	99.5
1-methoxy 2-propanol acetate	Kodak	99.9
2-(2-propoxyethoxy) ethanol	Kodak	99.0
2-(2-butoxyethoxy) ethanol	Kodak	99.0
monoethanolamine	Fisher	99.9
valeric acid	Aldrich	99.0
ethyl-3-ethoxy propionate	Kodak	99.9

Table 2. Critical temperature and pressures of substances studied

Substance	T_c(K) static	T_c(K) flow	P_c(MPa) flow
acetophenone	709.5 ± 0.4	709.6 ± 1.0	4.01 ± 0.03
2-butoxyethanol	633.9 ± 1.1	633.9 ± 0.6	3.27 ± 0.02
2-propoxyethanol	614.6 ± 0.3	615.2 ± 0.6	3.65 ± 0.02
ethylene glycol	718 ± 9	>720	>9
ethyl-3-ethoxy propionate	618.7 ± 1.8	621.0 ± 0.6	2.66 ± 0.02
1-methoxy 2-propanol acetate	597.6 ± 1.1	597.9 ± 0.6	3.01 ± 0.02
2-(2-propoxyethoxy) ethanol	687 ± 13	679.8 ± 1.0	3.00 ± 0.02
2-(2-butoxyethoxy) ethanol		692.3 ± 1.4	2.79 ± 0.07
monoethanolamine	>670	671.4 ± 1.4	8.03 ± 0.04
valeric acid	639.9 ± 0.4	637.2 ± 0.7	3.63 ± 0.03

DIPPR PROJECT 871: DETERMINATION OF IDEAL-GAS ENTHALPIES OF FORMATION FOR KEY COMPOUNDS: THE 1989 PROJECT RESULTS

W. V. Steele, R. D. Chirico, A. Nguyen, I. A. Hossenlopp, and N. K. Smith

■ IIT Research Institute, National Institute for Petroleum and Energy Research, P.O. Box 2128, Bartlesville, OK 74005

The results of a study aimed at improvement of group-contribution methodology for estimation of thermodynamic properties of organic and organosilicon substances are reported. Specific weaknesses where particular group-contribution terms were unknown, or estimated because of lack of experimental data, are addressed by experimental studies of enthalpies of combustion in the condensed phase, vapor-pressure measurements, and differential scanning calorimetric (d.s.c.) heat-capacity measurements. Ideal-gas enthalpies of formation of ($\pm$)-butan-2-ol, tetradecan-1-ol, hexan-1, 6-diol, methacrylamide, benzoyl formic acid, naphthalene-2, 6-dicarboxylic acid dimethyl ester, and tetraethylsilane are reported. A crystalline-phase enthalpy of formation at 298.15 K was determined for naphthalene-2, 6-dicarboxylic acid, which decomposed at 695 K before melting. The combustion calorimetry of tetraethylsilane used the proven fluorine-additivity methodology. Critical temperature and critical density were determined for tetraethylsilane with the d.s.c. and the critical pressure was derived. Group-additivity parameters useful in the application of group-contribution correlations are derived.

This research was funded jointly by the U. S. Department of Energy (DOE) through the Office of Fossil Energy's Advanced Exploratory Research program and the Design Institute for Physical Property Data (DIPPR) of the American Institute of Chemical Engineers through some of its member industrial organizations. The work performed in the third year of this project, (DIPPR Research Project 871: Determination of Pure Compound Ideal-Gas Enthalpies of Formation), represents the outcome of a meeting in late 1988, and subsequent communications, in which representatives of the DOE Bartlesville Project Office, the Design Institute for Physical Property Data (DIPPR), and the National Institute for Petroleum and Energy Research (NIPER) agreed on a list of compounds for which the determination of the enthalpy of formation in the ideal-gas state would be of benefit to all the participants.

Research programs funded by DOE Fossil Energy at NIPER share a common goal: the accurate estimation of both the thermochemical and thermophysical properties for a range of organic compounds which are important in the processing of alternate fuel sources. Our research has shown that there are a number of key "small" organic compounds for which thermochemical and thermophysical properties are incomplete, in question, or just completely unknown. Data on these compounds will greatly enhance the application of group-contribution methodology (1,2) as a property-estimation tool. In particular, the determination of the ideal-gas enthalpies of formation of a series of simple ring systems is a necessary forerunner to the development of a scheme for the accurate estimation of data for large molecules which contain these structural entities.

DIPPR's goal is to develop, organize, maintain, and make available reliable physical, thermodynamic, and transport property data for industrially important chemical compounds. Work is in progress compiling data on >1000 compounds important to industry. Where no data exist, estimation is attempted. These estimations require a strong base of accurate and precise data on basic molecular structures.

The evaluation of chemical plant safety has never been as important as it is today. The ideal-gas enthalpy of formation is the thermodynamic property most needed for evaluation of the energy hazard potential of an organic compound. A subcommittee of

ASTM E27 has written the computer program CHETAH (Chemical Thermodynamic and Energy Release Evaluation) which estimates gas-phase thermochemical data for organic materials using a second-order group-contribution method. The second-order group-contribution methodology for the calculation of thermodynamic properties has been outlined in detail by Benson (1). However, this text lacks parameters for a number of important groups and correction terms for several important ring structures. Parameters for some structural groups were derived from data which have since been shown to be incorrect. In the absence of data, application of the methodology for the estimation of thermochemical properties for some important organic compound types is impossible.

Whereas the condensed-phase enthalpy of formation of a compound is of greatest interest in the calculation of energy balances for a given chemical process, the enthalpy of formation for the ideal-gas state is of greatest interest in the general case, where the answer can be used to derive a group parameter or correction factor. In the latter case, this single value can give sufficient information to enable estimations for a large group of compounds containing that molecular entity.

In summary, the objective of this project is to expand the group-additivity method of calculation of thermodynamic properties by determining thermochemical data on compounds containing unique groups or atomic environments.

In the third year of the project, eight compounds were chosen for experimental studies. These compounds and their molecular structures are listed in Table 1. The derivation of ideal-gas standard enthalpies of formation for each of the compounds required experimental measurements in addition to the determination of the standard enthalpies of combustion. A listing of the required auxiliary measurements for each of the compounds is also given in Table 1.

The purity of the sample employed in a measurement of a thermodynamic property can significantly affect the accuracy of the measurement. The degree of inaccuracy introduced by the presence of impurities depends on a number of factors. In the case of the

Table 1. Outline of sample measurements performed in this project [a]

Compound	$\Delta_c U^o_m$	Vap. pressure	Heat capacity
(±)-Butan-2-ol	x	x	
Tetradecan-1-ol	x	x	x
Hexan-1,6-diol	x	x	x
Methacrylamide	x	x	x
Benzoyl formic acid	x	x	x
Naphthalene-2,6-dicarboxylic acid	x	[b]	x
Naphthalene-2,6-dicarboxylic acid dimethyl ester	x	x	x
Tetraethylsilane	x	x	x

[a] Measurements made are denoted by x
[b] Attempted but not successful. See text.

$$\text{CH}_3\text{CHCH}_2\text{CH}_3 \text{ (OH)} \quad \text{CH}_3(\text{CH}_2)_{12}\text{CH}_2\text{OH}$$

Butan-2-ol   Tetradecan-1-ol

$$\text{HOCH}_2(\text{CH}_2)_4\text{CH}_2\text{OH} \quad \text{CH}_2\text{CCONH}_2 \text{ (CH}_3)$$

Hexan-1,6-diol   Methacrylamide

Benzoyl formic acid

naphthalene-2,6-dicarboxylic acid

naphthalene-2,6-dicarboxylic acid dimethyl ester

$$(\text{CH}_3\text{CH}_2)_4\text{Si}$$

Tetraethylsilane

measurement of enthalpies of combustion, the presence of small amounts (less than 0.1 percent) of isomeric impurities usually will not have a significant effect on the result. However, this rule of thumb must be used with care, especially if the major impurity is an isomer with increased stability due to resonance or instability due to steric interactions.

EXPERIMENTAL

In this section, details are given of the apparatus and procedures used in obtaining the experimental data. These have been previously described in the literature and in various DOE reports. Therefore, details have been kept to a minimum here and the literature referenced for further consultation.

Materials

To minimize errors due to impurities, care was taken to ensure only samples of high purity (>99.9 mole percent purity) were subjected to the calorimetric measurements. With the exceptions of naphthalene-2,6-dicarboxylic acid and naphthalene-2,6-dicarboxylic acid dimethyl ester all compounds were purchased from Aldrich Chemical Company. Gas-liquid chromatographic (glc) analyses on the purchased samples gave an average purity of 99.8 mole percent. The compounds were purified by the Oklahoma State University Synthesis and Purification Group under the direction of Professor E. J. Eisenbraun. Glc analyses of the calorimetric samples gave purities of at least 99.95 mole percent for each compound. The samples of naphthalene-2,6-dicarboxylic acid and naphthalene-2,6-dicarboxylic acid dimethyl ester were supplied (purity >99.95 mole percent) by an industrial participant in the project. The high purity of each calorimetric sample was confirmed subsequently by the percentage CO_2 recoveries in the combustion calorimetric measurements, the small differences between the boiling and condensation temperatures in the ebulliometric vapor-pressure measurements, and in the enthalpy-of-fusion measurements in the d.s.c. studies (all of which are described in detail in this report).

All transfers of the calorimetric samples were done under nitrogen, helium, or by vacuum distillation. The water used as a reference material in the ebulliometric vapor-pressure measurements was deionized and distilled from potassium permanganate. The n-decane used as a reference material for the ebulliometric measurements was purified by urea complexation, two recrystallizations of the complex, decomposition of the complex with water, extraction with ether, drying with $MgSO_4$, and distillation at 337 K and 1 kPa pressure. Glc analysis of the n-decane sample failed to show any impurity peaks.

Physical Constants

Molar values are reported in terms of the 1981 relative atomic masses (3) and the gas constant, $R = 8.31451$ $J \cdot K^{-1} \cdot mol^{-1}$, adopted by CODATA (4). The platinum resistance thermometers used in these measurements were calibrated by comparison with standard thermometers whose constants were determined at the National Institute for Standards and Technology (NIST), formerly the National Bureau of Standards (NBS). All temperatures are reported in terms of the IPTS-68 (5). Measurements of mass, time, electric resistance, and potential difference were made in terms of standards traceable to calibrations at NIST.

Energy of Combustion Apparatus and Procedures

The apparatus and experimental procedures used in the combustion calorimetry of organic C,H,N,O compounds at the National Institute for Petroleum and Energy Research have been described (6-9). (The combustion calorimetry of tetraethylsilane is a special case and details follow in the next section.) A rotating-bomb calorimeter (laboratory designation BMR II) (10) and a platinum-lined bomb (laboratory designation Pt-3b) (11) with an internal volume of 0.3934 dm^3 were used without rotation. Each experiment was started at 296.15 K and, by judicious choice of sample and auxiliary masses, completed very close to 298.15 K. Flexible borosilicate-glass ampoules (6,12) were used to confine the butan-2-ol sample.

NBS benzoic acid (sample 39i) was used for calibration of the calorimeter; its specific energy of combustion is $-(26434.0\pm3.0)$ J·g^{-1} under certificate conditions. Conversion to standard states (13) gives $-(26413.7\pm3.0)$ J·g^{-1} for $\Delta_c U_m^o/M$, the standard specific energy of the idealized combustion reaction. The combustion measurements were performed in two separate series. Calibration experiments were interspersed with each series of measurements. Nitrogen oxides were not formed in the calibration experiments due to the high purity of the oxygen used and preliminary bomb flushing. The energy equivalent of the calorimeter, ε(calor), obtained for each calibration series was (16768.3 ± 0.4) J·K^{-1} (mean and standard deviation of the mean) for the butan-2-ol, tetradecan-1-ol, hexan-1,6-diol, methacrylamide, and benzoyl formic acid measurements, and (16768.4 ± 0.6) J·K^{-1} for the naphthalene-2,6-dicarboxylic acid and naphthalene-2,6-dicarboxylic acid dimethyl ester measurements.

The auxiliary oil (laboratory designation TKL66) had the empirical formula $CH_{1.913}$. For this material, $\Delta_c U_m^o/M$ was $-(46042.5\pm1.8)$ J·g^{-1} (mean and standard deviation). For the cotton fuse, empirical formula $CH_{1.774}O_{0.887}$, $\Delta_c U_m^o/M$ was -16945 J·g^{-1}. Information necessary for reducing apparent mass measured in air to mass, converting the energy of the actual bomb process to that of the isothermal process, and reducing to standard states (13) is given in Table 2. Values of density reported in Table 2 were measured in this laboratory, either from measurements of volumes of the ampoules used in the combustion calorimetry, and their enclosed sample masses, for the liquid, butan-2-ol, or from the dimensions of a pellet of known mass for the remaining compounds which were crystalline solids at 298.15 K. Values of the heat capacity of each sample at 298.15 K were measured using a differential scanning calorimeter as described later.

Table 2. Physical properties at 298.15 K [a,b]

Compound	ρ kg·m^{-3}	$10^7(\delta v/\delta T)_p$ m^3·K^{-1}	C_p/R
($\pm$)-Butan-2-ol	802.5	0.96	23.7
Tetradecan-1-ol	949.6	(0.3)	51.3
Hexan-1,6-diol	1079	(0.3)	24.4
Methacrylamide	1080	(0.3)	17.3
Benzoyl formic acid	1340	(0.3)	23.2
Naphthalene-2,6-dicarboxylic acid	1330	(0.3)	27.7
Naphthalene-2,6-dicarboxylic acid dimethyl ester	1321	(0.3)	23.1
Tetraethylsilane	768.3	2.0	35.8

[a] Values in parentheses are estimates.
[b] See text for details of the density and heat-capacity measurements (crystalline phase except ($\pm$)-butan-2-ol and tetraethylsilane both liquid phase).

Nitric acid, formed during the nitrogen-containing compound (methacrylamide) combustions, was determined by titration with standard sodium hydroxide (14). Carbon dioxide was also recovered from the combustion products of each experiment. Anhydrous lithium hydroxide was used as adsorbent (7). The combustion products were checked for unburned carbon and other products of incomplete combustion, but none was detected. Summaries of the carbon dioxide recoveries for each calibration series and the corresponding compound energy determinations are listed in Table 3.

Energy of Combustion of Tetraethylsilane

The rotating-bomb procedure for the combustion of silicon compounds in oxygen using a fluorine auxiliary to promote the combustion and form a well-defined final solution of fluorosilicic acid in excess hydrofluoric acid was first devised by Good et al. (15). The calorimetric procedure devised by Good et al. was used in the study of tetraethylsilane reported here. Benzotrifluoride was selected as the fluorine-containing solvent for tetraethylsilane because of availability and previous energy-of-combustion determination in the laboratory (15). To mix accurately weighed amounts of the two volatile liquids the tetraethylsilane was first sealed in a small container of polyester film that also contained a piece of platinum with a serrated edge. This container then was sealed with the benzotrifluoride in a larger container of polyester film. By weighing at appropriate stages, the masses of tetraethylsilane, benzotrifluoride, polyester film, and platinum were determined individually. The two liquids were mixed intimately by rupturing the inner container with the platinum and manipulating the outer container.

For the benzotrifluoride, molecular formula $C_7H_5F_3$, $\Delta_c U^o_m/M = -(23051.4 \pm 1.7) J \cdot g^{-1}$, was determined previously (15). The value for $\Delta_c U^o_m/M$ obtained for the polyester film, empirical formula $C_{10}H_8O_4$, was a function of the relative humidity (RH) in the laboratory during weighings (16):

$$\{(\Delta_c U^o_m/M)/(J \cdot g^{-1})\} = -22912.0 - 1.0560(RH). \quad (1)$$

Auxiliary information, necessary for reducing weights measured in air to masses, converting the energy of the actual bomb process to that of the isothermal process, and reducing to standard states (15,17), was taken from reference 15 for all the materials used except tetraethylsilane. For tetraethylsilane the values used are reported in Table 2.

Comparison experiments (17) were used to minimize errors from inexact reduction to standard states caused by a lack of values necessary to correct for such effects as the solubility and enthalpy of solution of CO_2 in solutions of HF and H_2SiF_6. In the comparison experiments the sample burned was the thermochemical standard benzoic acid. The amount of benzoic acid used was selected so that the energy evolved and the CO_2 produced in the comparison experiment were nearly the same as in the companion combustion experiment. The bomb initially contained an aqueous mixture of HF and H_2SiF_6, which upon dilution with the water formed by the combustion of the benzoic acid, gave a solution of nearly the same amount and concentration as in the combustion experiment.

Table 3. Carbon dioxide recoveries

Compound	No. of experiments	Percent recovery [a]
benzoic acid calibration	6	99.983±0.018
(±)-Butan-2-ol	6	99.994±0.016
Tetradecan-1-ol	6	99.986±0.024
Hexan-1,6-diol	7	100.021±0.027 [b]
Methacrylamide	6	99.810±0.028 [b]
Benzoyl formic acid	6	99.960±0.031
benzoic acid calibration	6	99.998±0.006
Naphthalene-2,6-dicarboxylic acid	7	99.996±0.008
Naphthalene-2,6-dicarboxylic acid dimethyl ester	7	99.942±0.013 [b]

[a] Mean and standard deviation of the mean.
[b] Results of combustion study based on percentage CO_2 recovery (See text).

Vapor Pressure Apparatus and Procedures

The essential features of the ebulliometric equipment and procedures for vapor-pressure measurements are described in the literature (18-20). The ebulliometers were used to reflux the substance under study with a standard of known vapor pressure under a common helium atmosphere. The boiling and condensation temperatures of the two substances were determined, and the vapor pressure was derived using the condensation temperature of the standard (20).

The precision in the temperature measurements for the ebulliometric vapor-pressure studies was 0.001 K. Uncertainties in the pressures are adequately described by:

$$\sigma(p) = (0.001) \{ (dp_{ref}/dT)^2 + (dp_\Xi/dT)^2 \}^{1/2}, \quad (2)$$

where p_{ref} is the vapor pressure of the reference substance and p_Ξ is the vapor pressure of the sample under study. Values of dp_{ref}/dT for the reference substances were calculated from fits of the Antoine equation (21) to vapor pressures of the reference materials (decane and water) reported in reference 20.

The equipment for the inclined piston vapor pressure measurements has been described by Douslin and McCullough, (22) and Douslin and Osborn (23). Recent revisions to the equipment and procedures have been reported (24). The low pressure range of the inclined-piston measurements, 10 to 3500 Pa, necessitated diligent outgassing of the sample prior to introduction into the apparatus. Also, prior to the sample introduction, all parts of the cell in contact with the sample were baked at 623 K under high vacuum (< 10^{-4} Pa). The thoroughly outgassed samples were placed in the apparatus, and additional outgassing was performed prior to commencing measurements. Finally, prior to each measurement, a small amount of sample was pumped off. Measurements were made as a function of time to extrapolate the pressure to the time when the pumping valve was closed; i.e., to the time when insignificant amounts of light gas had leaked into the system or diffused out of the sample.

Uncertainties in the pressures determined with the inclined-piston apparatus, on the basis of estimated precision of measuring the mass, area, and angle of inclination of the piston, are adequately described by the expression:

$$\sigma(p) = 1.5 \times 10^{-4} p + 0.2 \text{ Pa}. \quad (3)$$

The uncertainties in the temperatures are 0.001 K.

Differential Scanning Calorimetry

The technique and methodology used in the differential scanning calorimetric (d.s.c.) measurement has been outlined in references 25 through 27. The major difference between our measurement technique and that used by Mraw et al. is the substitution of specially designed cells (28) for the aluminum "volatile sample cells." These cells, designed and manufactured at NIPER, are made of 17-4 PH stainless steel and can withstand both high pressures (to 7.6 MPa) and high temperatures (to 900 K). The d.s.c. was used to determine the enthalpies of fusion and heat capacities of each compound (except butan-2-ol) over a range of temperature. In normal operation (28), correction is made for the enthalpy involved in the vaporization of small amounts of the sample under study into the vapor space of the sealed cells.

The theoretical background for the determination of heat capacities at vapor-saturation pressure, $C_{sat,m}$, with results obtained with a d.s.c. has been described (28,29). If two phases are present and the liquid is a pure substance, then the vapor pressure p and the chemical potential μ are independent of the amount of substance n and the cell volume V_x, and are equal to p_{sat} and μ_{sat}. The two-phase heat capacities at cell volume V_x, $C^{II}_{x,m}$, can be expressed in terms of the temperature derivatives of these quantities:

$$n C^{II}_{x,m}/T = -n(\delta^2\mu/\delta T^2)_{sat} + V_x(\delta^2 p/\delta T^2)_{sat} + \{(\delta V_x/\delta T)_x (\delta p/\delta T)_{sat}\}. \quad (4)$$

The third term on the right-hand side of equation (4) includes the thermal expansion of the cell. In this research the thermal expansion of the cells was expressed as:

$$V_x(T) / V_x(298.15 \text{ K}) = 1 + ay + by^2, \quad (5)$$

where, y = (T - 298.15) K, a = 3.216 x 10^{-5} K^{-1}, and b = 5.4 x 10^{-8} K^{-2}.

$(\delta p/dT)_{sat}$ can be calculated based on the vapor pressures measured in this research. Therefore, with a minimum of two different filling levels of the cell $(\delta^2 p/dT^2)_{sat}$ and $(\delta^2\mu/\delta T^2)_{sat}$ can be determined. In practice normally three cell fillings spanning a range of density (0.6 to 1.4 times ρ_c) are used. To obtain the saturation heat capacity $C_{sat,m}$ at vapor pressures greater than 0.1 MPa, the limit where the cell is full of liquid is required; i.e., $(n/V_x) = \{1/V_m(l)\}$ where $V_m(l)$ is the molar volume of the liquid:

$$\lim_{(n/V_x) \to \{1/V_m(l)\}} (n\, C^{II}_{V,m}/T) =$$
$$V_m(l)(\delta^2 p/\delta T^2)_{sat} - n(\delta^2 \mu/\delta T^2)_{sat}. \quad (6)$$

$C_{sat,m}$ is obtained using the expression:

$$\lim_{(n/V_x) \to \{1/V_m(l)\}} (n\, C^{II}_{V,m}) =$$
$$n[C_{sat,m} - \{T(\delta p/\delta T)_{sat}(dV_m(l)/dT)\}]. \quad (7)$$

Thus, reliable liquid density values are also required to determine $C_{sat,m}$.

RESULTS

A typical combustion experiment for each C,H,N,O compound studied is summarized in Table 4. It is impractical to list summaries for each combustion, but values of $\Delta_c U^o_m/M$ for all the experiments are reported in Table 5. Values of $\Delta_c U^o_m/M$ in Tables 4 and 5 for the C,H,N,O compounds refer to the general reaction:

$$C_aH_bN_cO_d\,(cr\ or\ l) + \left(a+\frac{b}{4}-\frac{d}{2}\right)O_2(g) =$$
$$aCO_2(g) + \frac{b}{2}H_2O(l) + \frac{c}{2}N_2(g). \quad (8)$$

For the compounds, hexan-1,6-diol, methacrylamide, and naphthalene-2,6-dicarboxylic acid the values of $\Delta_c U^o_m/M$ refer to unit mass of sample derived from the corresponding carbon dioxide analysis of the combustion products. Table 6 gives derived values of the standard molar energy of combustion $\Delta_c U^o_m$; the standard molar enthalpy of combustion $\Delta_c H^o_m$; and the standard molar enthalpy of formation $\Delta_f H^o_m$ for the compounds studied. Values of $\Delta_c U^o_m$ and $\Delta_c H^o_m$ for the C,H,N,O compounds refer to Equation 8. The corresponding values of $\Delta_f H^o_m$ refer to the reaction:

$$aC(cr, graphite) + \frac{b}{2}H_2(g) + \frac{c}{2}N_2(g) + \frac{d}{2}O_2(g) =$$
$$C_aH_bN_cO_d\,(cr\ or\ l). \quad (9)$$

Uncertainties given in Table 6 are the "uncertainty interval" (30). The enthalpies of formation of $CO_2(g)$ and $H_2O(l)$ were taken to be $-(393.51 \pm 0.13)$ and $-(285.830 \pm 0.042)$ kJ·mol^{-1}, respectively, as assigned by CODATA (31).

Table 4. Typical combustion experiments at 298.15 K (p^o = 101.325 kPa) [a,b]

	A	B
m'(compound)/g	0.872314	0.781654
m''(oil)/g	0.045762	0.0
m'''(fuse)/g	0.001652	0.001610
$n_i(H_2O)$/mol	0.05535	0.05535
m(Pt)/g	19.924	20.804
$\Delta T/K = (t_i - t_f + \Delta t_{corr})/K$	1.98964	1.98970
ε(calor)(ΔT)/J	-33362.9	-33363.9
ε(cont)(ΔT)/J [c]	-36.8	-36.6
ΔU_{ign}/J	0.7$_5$	0.7$_5$
ΔU(corr. to std. states)/J [d]	11.0	10.3
$-m''(\Delta_c U^o_m/M)$(oil)/J	2107.0	0.0
$-m'''(\Delta_c U^o_m/M)$(fuse)/J	28.0	27.3
$m'(\Delta_c U^o_m/M)$(compound)/J	-31252.9	-33362.1
$(\Delta_c U^o_m/M)$(compound)/J·g^{-1}	-35827.6	-42681.4

	C	D
m'(compound)/g	1.040842	1.216268
m''(fuse)/g	0.001712	0.001701
$n_i(H_2O)$/mol	0.05535	0.05535
m(Pt)/g	19.924	19.924
$\Delta T/K = (t_i - t_f + \Delta t_{corr})/K$	1.98162	1.98764
ε(calor)(ΔT)/J	-33228.5	-33329.3
ε(cont)(ΔT)/J [c]	-36.9	-37.2
ΔU_{ign}/J	0.7$_5$	0.7$_5$
ΔU(corr. to std. states)/J [d]	12.6	17.9
$\Delta U_{dec}(HNO_3)$/J	0.0	80.2
$-m''(\Delta_c U^o_m/M)$(fuse)/J	29.0	25.9
$m'(\Delta_c U^o_m/M)$(compound)/J	-33223.0	-33241.7
$(\Delta_c U^o_m/M)$(compound)/J·g^{-1}	-31919.3	-27331.0

	E	F
m'(compound)/g	1.390517	1.426682
m''(fuse)/g	0.001704	0.001797
$n_i(H_2O)$/mol	0.05535	0.05535
m(Pt)/g	20.804	19.926
$\Delta T/K = (t_i - t_f + \Delta t_{corr})/K$	1.94468	2.00666
ε(calor)(ΔT)/J	-32608.9	-33648.5
ε(cont)(ΔT)/J [c]	-37.5	-38.0
ΔU_{ign}/J	0.7$_5$	0.7$_5$
ΔU(corr. to std. states)/J [d]	29.2	31.4
$-m''(\Delta_c U^o_m/M)$(fuse)/J	28.9	30.5
$m'(\Delta_c U^o_m/M)$(compound)/J	-32587.5	-33623.8
$(\Delta_c U^o_m/M)$(compound)/J·g^{-1}	-23435.5	-23568.0

Table 4. Continued

	G
m'(compound)/g	1.261291
m''(fuse)/g	0.001705
$n_i(H_2O)$/mol	0.05535
$m(Pt)$/g	20.805
$\Delta T/K = (t_i - t_f + \Delta t_{corr})/K$	2.01072
$\varepsilon(calor)(\Delta T)/J$	-33716.6
$\varepsilon(cont)(\Delta T)/J$ [c]	-38.0
$\Delta U_{ign}/J$	0.7$_5$
ΔU(corr. to std. states)/J [d]	26.4
$-m''(\Delta_c U_m^o/M)$(fuse)/J	28.9
$m'(\Delta_c U_m^o/M)$(compound)/J	-33698.5
$(\Delta_c U_m^o/M)$(compound)/J·g^{-1}	-26717.6

[a] The symbols and abbreviations of this Table are those of reference 13 except as noted.
[b] **A** = (±)-butan-2-ol; **B** = tetradecan-1-ol; **C** = hexan-1,6-diol; **D** = methacrylamide; **E** = benzoyl formic acid; **F** = naphthalene-2,6-dicarboxylic acid; and **G** = naphthalene-2,6-dicarboxylic acid dimethyl ester.
[c] $\varepsilon_i(cont)(t_i - 298.15\ K) + \varepsilon_f(cont)(298.15\ K - t_f + \Delta t_{corr})$
[d] Items 81 to 85, 87 to 90, 93, and 94 of the computational form of reference 13.

Results from a typical combustion and its companion comparison experiment for tetraethylsilane are summarized in Table 4A. Six successful combustions and corresponding comparison experiments were made and the results are summarized in Table 5. The measured values for the energy of combustion and subsequent solution of the silicon dioxide or silicon tetrafluoride formed can, on average, be represented by the following equation:

$$C_8H_{20}Si\ (l) + 14\ O_2\ (g) + 15.15_5\ HF \cdot 413 H_2O \rightarrow$$
$$8 CO_2\ (g) + H_2SiF_6 \cdot 9.15_5 HF \cdot 425 H_2O. \quad (10)$$

The thermochemical cycle used to calculate the enthalpy of formation of tetraethylsilane is given in Table 6A. The footnote to Table 6A lists the sources of the auxiliary thermochemical values (31-33) required to determine the values reported in Table 6.

Table 4A. Typical combustion and comparison experiments for tetraethylsilane (Et$_4$Si) at 298.15 K and 101.325 kPa [a]

	comb.	**compar.**
m(tetraethylsilane)/g	0.201115	
m(benzotrifluoride)/g	0.993319	
m(polyester film)/g	0.101393	
m(benzoic acid)/g		1.301002
m(fuse)/g	0.001335	0.001753
$n_i(H_2O)$/mol	0.05535	0.052693
$n_i(HF)$/mol		0.011685
$n_i(H_2SiF_6)$/mol		0.001345
$m(Pt)$/g	20.805	20.805
$\Delta T/K = (t_i - t_f + \Delta t_{corr})/K$	2.05674	2.04660
$\varepsilon(calor)(\Delta T)/J$	-34530.6	-34362.0 [b]
$\varepsilon(cont)(\Delta T)/J$ [c]	-114.7	-111.6
$\Delta U_{ign}/J$	0.7$_5$	0.7$_5$
ΔU(corr. to std. states)/J [d]	41.6	51.1
$-m(\Delta_c U_m^o/M)$(fuse)/J	22.6	29.7
$-m(\Delta_c U_m^o/M)$(benzoic acid)/J		-34392.0
$-m(\Delta_c U_m^o/M)$(polyester)/J	-2328.5	
$-m(\Delta_c U_m^o/M)$(benzotrifluoride)/J	-22898.8	
$\Delta U_{sol}/J$ [e]	3.2	
$m'(\Delta_c U_m^o/M)$(Et$_4$Si)/J	-9349.8	
$(\Delta_c U_m^o/M)$(Et$_4$Si)/J·g^{-1}	-46490	

[a] The symbols and abbreviations of this Table are those of references 13, 15 and 17 except as noted.
[b] Value used to determine ε(calor) for the corresponding combustion experiment (see text).
[c] $\varepsilon_i(cont)(t_i - 298.15\ K) + \varepsilon_f(cont)(298.15\ K - t_f + \Delta t_{corr})$
[d] Items 81 to 85, 87 to 90, 93, and 94 of the computational form of references 13 and 17; correction to standard states.
[e] Thermochemical correction for the solution of tetraethylsilane in benzotrifluoride: solution of 1 mole of tetraethylsilane in 3 moles of benzotrifluoride was accompanied by the adsorption of 2316 J.

Table 5. Summary of experimental energy of combustion results. T = 298.15 K, and p° =101.325 kPa. The uncertainties shown are one standard deviation of the mean.

(±)-Butan-2-ol

$\{(\Delta_c U_m^o/M)\text{(compound)}\}/(J \cdot g^{-1})$

-35827.6   -35828.5   -35823.2   -35829.9
            -35830.5   -35824.6

$\{(\Delta_c U_m^o/M)\text{(compound)}\}/(J \cdot g^{-1}) > -35827.4 \pm 1.2$

Tetradecan-1-ol

$\{(\Delta_c U_m^o/M)\text{(compound)}\}/(J \cdot g^{-1})$

-42681.4   -42684.0   -42684.4   -42684.6
           -42684.9   -42681.7

$\{(\Delta_c U_m^o/M)\text{(compound)}\}/(J \cdot g^{-1}) > -42683.5 \pm 0.6$

Table 5. Continued

Hexan-1,6-diol

$\{(\Delta_c U_m^o/M)(\text{compound})\}/(\text{J}\cdot\text{g}^{-1})$

-31919.3 -31917.9 -31910.8 -31915.3
-31924.2 -31914.4 -31914.8

$\{(\Delta_c U_m^o/M)(\text{compound})\}/(\text{J}\cdot\text{g}^{-1})>$ -31916.7±1.6

Methacrylamide

$\{(\Delta_c U_m^o/M)(\text{compound})\}/(\text{J}\cdot\text{g}^{-1})$

-27331.0 -27325.0 -27326.2 -27318.7
-27321.9 -27305.2 [a]

$\{(\Delta_c U_m^o/M)(\text{compound})\}/(\text{J}\cdot\text{g}^{-1})>$ -27324.7±2.1

Benzoyl formic acid

$\{(\Delta_c U_m^o/M)(\text{compound})\}/(\text{J}\cdot\text{g}^{-1})$

-23435.5 -23429.9 -23429.7 -23435.1
-23441.2 -23436.3

$\{(\Delta_c U_m^o/M)(\text{compound})\}/(\text{J}\cdot\text{g}^{-1})>$ -23434.7±1.8

Naphthalene-2,6-dicarboxylic acid

$\{(\Delta_c U_m^o/M)(\text{compound})\}/(\text{J}\cdot\text{g}^{-1})$

-23568.0 -23564.5 -23566.7 -23568.5
-23568.8 -23569.8

$\{(\Delta_c U_m^o/M)(\text{compound})\}/(\text{J}\cdot\text{g}^{-1})>$ -23567.7±0.8

Naphthalene-2,6-dicarboxylic acid dimethyl ester

$\{(\Delta_c U_m^o/M)(\text{compound})\}/(\text{J}\cdot\text{g}^{-1})$

-26717.6 -26708.9 -26716.8 -26717.0
-26718.4 -26712.9

$\{(\Delta_c U_m^o/M)(\text{compound})\}/(\text{J}\cdot\text{g}^{-1})>$ -26715.2±1.5

Tetraethylsilane [b]

$\{(\Delta_c U_m^o/M)(\text{compound})\}/(\text{J}\cdot\text{g}^{-1})$

-46490 -46558 -46443 -46467
-46492 -46495

$\{(\Delta_c U_m^o/M)(\text{compound})\}/(\text{J}\cdot\text{g}^{-1})>$ -46491±16

[a] Value excluded from average (see text).
[b] Value for the following reaction (see text):
$C_8H_{20}Si$ (l) + 14 O_2 (g) + 15.155HF·413H$_2$O
$\rightarrow$ 8CO$_2$ (g) + H$_2$SiF$_6$·9.155HF·425H$_2$O.

Table 6. Condensed phase molar thermochemical functions at 298.15 K and p^o = 101.325 kPa.[a]

	$\Delta_c U_m^o$/kJ·mol^{-1}	$\Delta_c H_m^o$/kJ·mol^{-1}	$\Delta_f H_m^o$/kJ·mol^{-1}
A	-2655.63±0.18	-2660.59±0.18	-342.60±0.25
B	-9151.06±1.16	-9168.41±1.16	-628.18±1.46
C	-3771.81±0.60	-3778.01±0.60	-583.86±0.72
D	-2325.49±0.46	-2327.35±0.46	-247.10±0.50
E	-3518.35±0.70	-3518.35±0.70	-487.22±0.80
F	-5095.19±0.78	-5095.19±0.78	-770.25±0.96
G	-6525.14±1.16	-6527.61±1.16	-696.51±1.34
H	-6710.2±4.6 [b]	-6725.1±4.6 [b]	-328.6±5.2

[a] A = (±)-Butan-2-ol; B = Tetradecan-1-ol; C = Hexan-1,6-diol; D = Methacrylamide; E = Benzoyl formic acid; F = Naphthalene-2,6-dicarboxylic acid; G = Naphthalene-2,6-dicarboxylic acid dimethyl ester; and H = tetraethylsilane.
[b] Value for the following reaction (see text):
$C_8H_{20}Si$ (l) + 14 O_2 (g) + 15.155HF·413H$_2$O
$\rightarrow$ 8CO$_2$ (g) + H$_2$SiF$_6$·9.155HF·425H$_2$O.
[c] Value for the following reaction (see text):
8C(c, graphite) + 10H$_2$(g) + Si(c) $\rightarrow$ $C_8H_{20}Si$ (l)

TABLE 6A Thermochemical cycle for tetraethylsilane

(I) 8CO$_2$ (g) + H$_2$SiF$_6$·9.155HF·425H$_2$O $\rightarrow$
$C_8H_{20}Si$ (l) + 14 O$_2$ (g) + 15.155HF·413H$_2$O
$\Delta_r H_m^o$ = (6725.1±4.6) kJ·mol^{-1}

(II) 8C(c, graphite) + 8O$_2$(g) $\rightarrow$ 8CO$_2$ (g)
$\Delta_r H_m^o$ = -(3148.08±1.04) kJ·mol^{-1}

(III) 10H$_2$(g) + 5O$_2$(g) $\rightarrow$ 10H$_2$O (l)
$\Delta_r H_m^o$ = -(2858.3±0.4) kJ·mol^{-1}

(IV) Si(c) + O$_2$(g) $\rightarrow$ SiO$_2$ (c, quartz)
$\Delta_r H_m^o$ = -(910.7±0.8) kJ·mol^{-1}

(V) SiO$_2$ (c, quartz) + 15.155HF·423H$_2$O $\rightarrow$
H$_2$SiF$_6$·9.155HF·425H$_2$O
$\Delta_r H_m^o$ = -(136.6±1.0) kJ·mol^{-1}

(VI) 15.155HF·413H$_2$O + 10H$_2$O (l) $\rightarrow$ 15.155HF·425H$_2$O
$\Delta_r H_m^o$ = -(0.005±0.001) kJ·mol^{-1}

(VII) 8C(c, graphite) + 10H$_2$(g) + Si(c) $\rightarrow$
$C_8H_{20}Si$ (l)
$\Delta_f H_m^o$ = -(328.6±5.2) kJ·mol^{-1}

[a] Enthalpies of reaction; (I) this research, (II) reference 31, (III) reference 31, (IV) reference 31, (V) reference 32, (VI) reference 33, and (VII) is the sum of reactions (I) through (VI).

Measured vapor pressures for butan-2-ol; tetradecan-1-ol; hexan-1,6-diol; benzoyl formic acid;

naphthalene-2,6-dicarboxylic acid dimethyl ester; and tetraethylsilane are listed in Table 7. Following previous practice (19), the results obtained in the ebulliometric measurements were adjusted to common pressures. The common pressures, the condensation temperatures, and the difference between the condensation and boiling temperatures for the samples are reported. The small differences between the boiling and condensation temperatures in the ebulliometric measurements indicated correct operation of the equipment and the high purity of the samples studied. Inclined-piston vapor-pressure measurements for naphthalene-2,6-dicarboxylic acid dimethyl ester and benzoyl formic acid are also listed in Table 7. For the ester the inclined-piston values extend the range of measured values down to 465 K. For benzoyl formic acid, sample decomposition at 429 K, (p>3.2 kPa), (see below) prevented determination of ebulliometric vapor pressures. Liquid-phase inclined-piston vapor-pressure measurements are reported in the temperature range 338.9 K (melting point) to 420 K.

The difference between the boiling and condensation temperatures (ΔT) for tetradecan-1-ol increased significantly above 569 K (see Table 7). An attempt was made to make a measurement at 169 kPa (593.5 K), but ΔT started at approximately 0.06 K and rapidly increased by several tenths of a degree. This phenomenon is indicative of sample decomposition. Similar behavior was observed for naphthalene-2,6-dicarboxylic acid dimethyl ester, where at 641.4 K (84 kPa) ΔT started at approximately 0.08 K and rapidly increased to 0.5 K.

All attempts to measure the vapor pressure of naphthalene-2,6-dicarboxylic acid in the solid phase were unsuccessful. At 523 K (the upper temperature limit of the apparatus at present) the vapor pressure was below the lower detection limit of the inclined piston, 10 Pa. The sample decomposed with a large pressure build-up before melting (695 K) preventing any attempt at obtaining liquid-phase measurements.

Table 7. Summary of vapor pressure results: decane or water refers to which material was used as the standard in the reference ebulliometer, the pressure p was calculated from the condensation temperature of the reference substance, T is the condensation temperature of the sample, $\Delta T = T_{boil} - T_{cond}$ is the difference between the boiling and condensation temperatures for the sample, Dp is the difference of the calculated value of pressure from the observed value of pressure, $\sigma(p)$ is the propagated error calculated from Equation (1).

Material	$\dfrac{T}{K}$	$\dfrac{p}{kPa}$	$\dfrac{\Delta p}{kPa}$	$\dfrac{\sigma(p)}{kPa}$	$\dfrac{\Delta T}{K}$
(±)-Butan-2-ol					
decane	301.900	3.0000	-0.0001	0.0002	0.033
decane	311.023	5.3330	0.0007	0.0004	0.028
decane	317.901	7.9989	-0.0004	0.0006	0.026
decane	323.029	10.666	-0.001	0.001	0.022
decane	327.157	13.332	0.000	0.001	0.014
decane	331.429	16.665	0.001	0.001	0.014
decane	334.969	19.933	-0.001	0.001	0.011
decane	339.612	25.023	0.001	0.002	0.008
water	339.615[a]	25.023	-0.003	0.002	0.003
water	344.275	31.177	-0.003	0.002	0.003
water	348.949	38.565	0.001	0.002	0.002
water	353.644	47.375	0.002	0.003	0.002
water	358.361	57.817	0.002	0.003	0.003
water	363.101	70.120	0.002	0.004	0.002
water	367.867	84.533	0.000	0.005	0.002
water	372.666	101.325	-0.023	0.006	0.002
water	377.484	120.79	0.00	0.01	0.004
water	382.343	143.25	-0.01	0.01	0.006
water	387.232	169.02	0.01	0.01	0.010
Tetradecan-1-ol					
decane	443.455	2.0000	-0.0003	0.0001	0.098
decane	450.026	2.6660	0.0001	0.0002	0.070
decane	459.817	3.9999	0.0006	0.0002	0.033
decane	467.161	5.3330	0.0002	0.0003	0.025
decane	478.124	7.9989	-0.0002	0.0004	0.018
decane	486.379	10.666	0.000	0.001	0.020
decane	493.079	13.332	-0.001	0.001	0.016
decane	500.053	16.665	-0.001	0.001	0.014
decane	505.863	19.933	-0.002	0.001	0.011
decane	513.524	25.023	0.000	0.001	0.008
water	513.521[a]	25.023	0.001	0.001	0.013
water	521.256	31.177	0.000	0.002	0.008
water	529.054	38.565	0.002	0.002	0.004
water	536.917	47.375	0.003	0.002	0.006
water	544.843	57.817	0.003	0.003	0.010
water	552.831	70.120	0.003	0.003	0.012
water	560.879	84.533	-0.006	0.004	0.015
water	568.969	101.325	0.003	0.004	0.031
water	577.096[a]	120.79	0.04	0.01	0.043
water	585.310[a]	143.25	-0.04	0.01	0.057
Hexan-1,6-diol					
decane	422.184	2.0000	0.0001	0.0001	0.124
decane	427.801	2.6660	0.0004	0.0002	0.076
decane	436.124	3.9999	-0.0003	0.0003	0.054
decane	442.321	5.3330	0.0000	0.0003	0.053
decane	451.520	7.9989	0.000	0.001	0.024
decane	458.412	10.666	-0.002	0.001	0.022
decane	463.974	13.332	-0.003	0.001	0.020
decane	469.741	16.665	-0.001	0.001	0.018

Table 7. Continued.

Material	$\frac{T}{K}$	$\frac{p}{kPa}$	$\frac{\Delta p}{kPa}$	$\frac{\sigma(p)}{kPa}$	$\frac{\Delta T}{K}$
\multicolumn{6}{c}{**Hexan-1,6-diol (cont)**}					
decane	474.530	19.933	0.000	0.001	0.017
decane	480.826	25.023	0.003	0.001	0.017
water	480.825 [a]	25.023	0.003	0.001	0.017
water	487.161	31.177	0.004	0.002	0.019
water	493.533	38.565	0.005	0.002	0.022
water	499.942	47.375	0.006	0.002	0.021
water	506.391	57.817	0.007	0.003	0.021
water	512.882	70.120	0.006	0.003	0.021
water	519.417	84.533	−0.003	0.004	0.015
water	525.994	101.325	−0.012	0.004	0.013
water	532.613	120.79	−0.02	0.01	0.014
water	539.277	143.25	−0.02	0.01	0.014
water	545.982	169.02	−0.03	0.01	0.011
water	552.729	198.49	0.01	0.02	0.012
water	559.524	232.02	0.03	0.02	0.013
\multicolumn{6}{c}{**Benzoyl formic acid**}					
IP	338.900	0.0147	0.0000	0.0002	
IP	340.000	0.0161	0.0000	0.0002	
IP	350.000	0.0343	0.0000	0.0002	
IP	360.000	0.0696	0.0000	0.0002	
IP	370.000	0.1350	0.0000	0.0002	
IP	380.000	0.2508	0.0000	0.0002	
IP	390.000	0.4484	0.0000	0.0003	
IP	400.000	0.7735	0.0000	0.0003	
IP	410.000	1.2912	0.0000	0.0004	
IP	420.000	2.0909	0.0000	0.0005	
\multicolumn{6}{c}{**Naphthalene-2,6-dicarboxylic acid dimethyl ester**}					
IP	465.011	0.3853	0.0008	0.0003	
IP	469.998	0.4817	0.0002	0.0003	
IP	475.003	0.5992	−0.0005	0.0003	
IP	479.998	0.7428	0.0002	0.0003	
IP	485.003	0.9139	−0.0008	0.0003	
IP	494.998	1.3638	−0.0011	0.0004	
IP	504.999	1.9967	0.0002	0.0005	
IP	510.002	2.3993	0.0014	0.0006	
IP	515.002	2.8676	0.0011	0.0006	
decane	546.383	7.9989	−0.0001	0.0004	0.031
decane	556.096	10.666	−0.001	0.001	0.022
decane	563.933	13.332	−0.001	0.001	0.023
decane	572.051	16.665	0.004	0.001	0.036
decane	578.802	19.933	0.000	0.001	0.029
water	587.668	25.023	−0.001	0.001	0.029
water	596.565	31.177	0.006	0.001	0.043
water	605.522	38.565	−0.002	0.002	0.040
water	614.510	47.375	−0.004	0.002	0.049
water	623.543	57.817	−0.010	0.003	0.056
water	632.578	70.120	0.010	0.003	0.083
\multicolumn{6}{c}{**Tetraethylsilane**}					
decane	318.628	2.0000	−0.0013	0.0001	0.076
decane	324.312	2.6660	−0.0001	0.0002	0.078
decane	332.782	3.9999	0.0010	0.0003	0.059
decane	339.119	5.3330	0.0027	0.0003	0.059
decane	348.579	7.9989	0.0027	0.0005	0.052
decane	355.694	10.666	0.002	0.001	0.048
decane	361.459	13.332	0.001	0.001	0.044
decane	367.458	16.665	−0.002	0.001	0.040
decane	372.449	19.933	−0.005	0.001	0.038
decane	379.027	25.023	−0.008	0.001	0.038
water	379.025 [a]	25.023	−0.006	0.001	0.036
water	385.656	31.177	−0.008	0.002	0.036
water	392.339	38.565	−0.008	0.002	0.035

Table 7. Continued.

Material	$\frac{T}{K}$	$\frac{p}{kPa}$	$\frac{\Delta p}{kPa}$	$\frac{\sigma(p)}{kPa}$	$\frac{\Delta T}{K}$
\multicolumn{6}{c}{**Tetraethylsilane (cont.)**}					
water	399.074	47.375	−0.006	0.002	0.036
water	405.863	57.817	−0.002	0.003	0.039
water	412.698	70.120	0.021	0.003	0.047
water	419.573 [a]	84.533	0.083	0.004	0.047
water	426.555	101.325	0.023	0.004	0.047
water	433.559	120.79	0.03	0.01	0.053
water	440.623	143.25	0.03	0.01	0.056
water	447.734	169.02	0.01	0.01	0.063
water	454.913	198.49	−0.06	0.01	0.061

[a] Point excluded from Cox equation fitting.

Previous studies by Scott and Osborn (34) have shown that the Cox equation (35) can adequately represent measured vapor pressures from the triple-point pressure to 0.3 MPa. The Cox equation in the form:

$$\ln(p/p_{ref}) = \{1 - (T_{ref}/T)\}\exp\{A + B(T/K) + C(T/K)^2\}, \quad (11)$$

was fit to the experimental vapor pressures with p_{ref} being chosen to be 101.325 kPa so that T_{ref} was the normal boiling temperature. In those fits, the sums of the weighted squares in the following function were minimized:

$$\Delta = \ln\{\ln(p/p_{ref})/(1 - T_{ref}/T)\} - A - B(T/K) - C(T/K)^2. \quad (12)$$

The weighting factors W are the reciprocals of the variance in Δ derived from the propagation of errors in the temperature and pressure determinations. W is defined as:

$$W = [(\delta\Delta/\delta T)_p^2 \{\sigma(T)\}^2 + (\delta\Delta/\delta p)_T^2 \{\sigma(p)\}^2]^{-1}. \quad (13)$$

Parameters derived from the fits are given in Table 8. For benzoyl formic acid the normal boiling point was fixed at 532.0 K. Details of the Cox equation fits are given in Table 7.

Enthalpies of vaporization $\Delta_l^g H_m$ were derived from the Cox equation fits using the Clapeyron equation:

$$dp/dT = \Delta_l^g H_m / (T\Delta_l^g V_m), \quad (14)$$

where $\Delta_l^g V_m$ is the increase in molar volume from the liquid to the real vapor. Estimates of second virial coefficients were made with the extended corresponding-states equation of Pitzer and Curl (36). Liquid-phase densities were also derived from corresponding-states using the formulation of Hales and Townsend (37):

$$(\rho/\rho_C) = 1.0 + 0.85\{1.0 - (T/T_C)\} + (1.692 + 0.986\omega)\{1.0 - (T/T_C)\}^{1/3}. \quad (15)$$

Third virial coefficients were estimated with the corresponding-states method of Orbey and Vera (38). This formulation for the third virial coefficient was applied successfully in analyses of the thermodynamic properties of benzene, toluene, and decane (39). The third virial coefficient is required for accurate calculation of the gas volume for pressures greater than one bar. Parameters (40,41) used to estimate the second and third virial coefficients for each compound are listed in Table 9A. Derived enthalpies of vaporization and entropies of compression are reported in Table 9.

Table 8. Cox equation coefficients. [a]

	A	B	C
T_{ref}/K	372.660	568.970	525.990
P_{ref}/kPa	101.325	101.325	101.325
A	3.05809	3.67463	3.62713
10^3B	-1.14891	-3.55062	-2.76856
10^6C	-0.02270	2.69238	1.75082
Range/K [b]	301 to 387	443 to 585	422 to 559

	E	G	H
T_{ref}/K	532.0 [c]	650.86	426.564
P_{ref}/kPa	101.325	101.325	101.325
A	3.08308	3.12543	2.84003
10^3B	-1.18877	-1.43697	-1.71981
10^6C	0.53278	0.82334	1.5532
Range/K [b]	339 to 420	465 to 633	318 to 455

[a] A = (±)-butan-2-ol; B = tetradecan-1-ol; C = hexan-1,6-diol: E = benzoyl formic acid G = naphthalene-2,6-dicarboxylic acid dimethyl ester and H = tetraethylsilane.
[b] Temperature range of the vapor pressures used in the fit.
[c] Temperature fixed at 532.0 K in fit.

Table 9. Enthalpies of vaporization obtained from the Cox and Clapeyron equations. [a]

T/K	$\Delta_l^g H_m$ / kJ·mol^{-1}	T/K	$\Delta_l^g H_m$ / kJ·mol^{-1}
(±)-Butan-2-ol			
298.15 [b]	49.86±0.03	360.00	42.84±0.23
300.00 [b]	49.68±0.03	380.00 [b]	39.96±0.38
320.00	47.65±0.07	400.00 [b]	36.72±0.57
340.00	45.39±0.12	420.00 [b]	33.17±0.78
Tetradecan-1-ol			
298.15 [b]	104.92±1.85	520.00	61.04±0.35
400.00 [b]	80.81±0.08	540.00	58.32±0.50
420.00 [b]	76.97±0.05	560.00	55.68±0.70
440.00	73.38±0.05	580.00	53.10±0.91
460.00	70.02±0.08	600.00 [b]	50.49±1.18
480.00	66.87±0.15	620.00 [b]	47.82±1.50
500.00	63.88±0.23		
Hexan-1,6-diol			
298.15 [b]	102.92±1.51	480.00	66.98±0.22
360.00 [b]	89.21±0.28	500.00	63.53±0.35
380.00 [b]	85.14±0.13	520.00	60.06±0.52
400.00 [b]	81.26±0.07	540.00	56.51±0.73
420.00	77.53±0.05	560.00	52.89±0.98
440.00	73.94±0.07	580.00 [b]	49.16±1.26
460.00	70.44±0.13	600.00 [b]	45.29±1.58
Benzoyl formic acid			
298.15 [b]	79.12±0.08	380.00	71.98±0.02
300.00 [b]	78.96±0.08	400.00	70.21±0.02
320.00 [b]	77.23±0.05	420.00	68.41±0.03
340.00	75.49±0.02	440.00 [b]	66.57±0.07
360.00	73.73±0.02	460.00 [b]	64.65±0.12
Naphthalene-2,6-dicarboxylic acid dimethyl ester			
298.15 [b]	99.67±5.15	580.00	71.27±0.57
460.00 [b]	82.61±0.15	600.00	69.31±0.80
480.00	80.69±0.10	620.00	67.25±1.10
500.00	78.80±0.10	640.00 [b]	65.06±1.48
520.00	76.93±0.17	660.00 [b]	62.73±1.96
540.00	75.07±0.27	680.00 [b]	60.21±2.56
560.00	73.19±0.40		
Tetraethylsilane			
298.15 [b]	44.62±0.05	400.00	38.11±0.27
300.00 [b]	44.51±0.05	420.00	36.70±0.37
320.00	43.26±0.03	440.00	35.19±0.52
340.00	42.01±0.07	460.00 [b]	33.56±0.68
360.00	40.76±0.10	480.00 [b]	31.79±0.88
380.00	39.46±0.17	500.00 [b]	29.87±1.10

[a] Uncertainty intervals are twice the standard deviation of the mean.
[b] Values at this temperature were calculated with extrapolated vapor pressures derived from the fitted Cox coefficients.

TABLE 9A. Critical constants.[a]

Compound		T_C K	p_C kPa	ρ_C (kg·m^{-3})
(±)-Butan-2-ol	[b]	535.95	4194	275.5
Tetradecan-1-ol	[c]	747	1810	232
Hexan-1,6-diol	[d]	700	2400	294
Benzoyl formic acid	[d]	743	3160	300
Naphthalene-2,6-dicarboxylic acid dimethyl ester		883	2485	350
Tetraethylsilane	[e]	606	2400	246

[a] Values used in the derivation of the enthalpies of vaporization listed in Table 9. See text.
[b] Reference 40
[c] Reference 41
[d] Estimated using unpublished group-additivity procedures developed in a research program at NIPER funded by DOE Office of Energy Research.
[e] This research.

Table 10 lists the experimental two-phase heat capacities $C_{x,m}^{II}$ determined by d.s.c. for tetradecan-1-ol, hexan-1,6-diol, benzoyl formic acid, naphthalene-2,6-dicarboxylic acid, naphthalene-2,6-dicarboxylic acid dimethyl ester, and tetraethylsilane obtained for the given cell fillings. Heat-capacities were determined at 20-K intervals with a heating rate of 0.083 K·s^{-1} and a 120 s equilibration period between heats. For each compound the upper temperature bound of the measurements was set by sample decomposition.

TABLE 10. Condensed-phase heat capacities and enthalpies of fusion.[a,b] (R=8.31451 J·K^{-1}·mol^{-1})

T/K	$C_{x,m}^{II}/R$	$C_{x,m}^{II}/R$	Phase
mass / g	0.015414	0.024312	
V. cell (cm^3)	0.05292	0.05288	

Tetradecan-1-ol

T/K	$C_{x,m}^{II}/R$	$C_{x,m}^{II}/R$	Phase
274.0	42.6	42.8	cr
284.0	45.6	46.3	cr
294.0	49.6	50.2	cr
324.0	64.9	65.3	l
334.0	66.5	67.1	l
344.0	68.2	68.7	l
354.0	70.0	70.5	l
364.0	71.4	72.0	l
374.0	72.7	73.2	l
384.0	74.0	74.4	l
394.0	74.9	75.2	l
404.0	75.5	75.0	l
414.0	74.8	76.2	l
424.0	76.0	76.6	l
434.0	76.2	76.7	l
444.0	76.2	76.8	l
454.0	76.3	77.0	l
464.0	76.3	77.2	l
474.0	76.6	77.2	l
484.0	76.8	77.4	l
494.0	77.3	77.7	l
504.0	77.8	77.7	l
514.0	78.1	78.1	l
524.0	78.9	78.7	l
534.0	79.4	80.1	l
544.0	81.1	79.7	l
554.0	81.1	80.4	l
564.0	81.9	80.9	l

Crystalline $C_{sat,m}/R = 0.360\,T - 56.0_6$ (in temperature range 269 to 311 K)
Liquid $C_{sat,m}/R = 0.159\,T + 13.61_8$ (in temperature range 311 to 379 K)

$\Delta_c^l H_m^o(311\ K) = 49.4\pm0.4$ kJ mol^{-1}

$\Delta_c^l H_m^o(298.15\ K) = 48.5\pm1.0$ kJ mol^{-1}

Hexan-1,6-diol

T/K	$C_{x,m}^{II}/R$	$C_{x,m}^{II}/R$	$C_{x,m}^{II}/R$	Phase
mass / g	0.017145	0.009932	0.016104	
V. cell (cm^3)	0.05288	0.05292	0.05292	
273.0	21.4	21.0		cr
283.0	22.4	22.6		cr
293.0	24.0	23.7		cr
303.0	25.2	25.0		cr
333.0	35.2	35.5		l
353.0	37.5	38.0		l
373.0	40.1	40.2		l
393.0	42.3	42.5		l
413.0	44.1	44.6	44.3	l
433.0	45.9	46.3	45.8	l
453.0	46.8	47.1	47.1	l
473.0	47.9	48.2	47.8	l
493.0	48.4		48.8	l
513.0	49.1		49.4	l
533.0	49.5		49.7	l
553.0	49.5		50.1	l
573.0			50.7	l
593.0			51.1	l
613.0			51.8	l
633.0			52.6	l
653.0			54.2	l
673.0			54.9	l
693.0			56.0	l

Crystalline $C_{sat,m}/R = 0.128\,T - 13.72$ (in temperature range 268 to 315 K)
Liquid $C_{sat,m}/R = 0.1123\,T - 1.91$ (in temperature range 315 to 413 K)

$\Delta_c^l H_m^o(315\ K) = 22.6\pm0.6$ kJ mol^{-1}

$\Delta_c^l H_m^o(298.15\ K) = 21.6\pm1.0$ kJ mol^{-1}

TABLE 10. Continued.

T/K	$C^{II}_{x,m}/R$	$C^{II}_{x,m}/R$	Phase
mass / g	0.013491	0.026761	
V. cell (cm^3)	0.05292	0.05282	

Benzoyl formic acid

T/K	$C^{II}_{x,m}/R$	$C^{II}_{x,m}/R$	Phase
295.0	23.1	23.2	cr
315.0	23.6	23.7	cr
355.0	32.4	32.2	l
375.0	33.2	32.9	l
395.0	33.9	33.8	l
415.0	34.9	34.6	l

Crystalline $C_{sat,m}/R = 0.025\,T + 15.7_8$ (in temperature range 285 to 338.9 K)
Liquid $C_{sat,m}/R = 0.0411\,T + 17.6_7$ (in temperature range 338.9 to 425 K)

$\Delta^l_c H^o_m(338.9\text{ K}) = 21.8 \pm 0.3$ kJ mol^{-1}

$\Delta^l_c H^o_m(298.15\text{ K}) = 19.4 \pm 0.6$ kJ mol^{-1}

Naphthalene-2,6-dicarboxylic acid dimethyl ester

T/K	$C^{II}_{x,m}/R$	$C^{II}_{x,m}/R$	Phase
mass / g	0.013392	0.018834	
V. cell (cm^3)	0.05292	0.05288	
320.00	36.3	36.9	cr
340.00	38.3	38.2	cr
360.00	39.7	39.7	cr
380.00	42.1	42.3	cr
400.00	43.9	43.4	cr
420.00	45.9	45.1	cr
440.00	47.7	48.9	cr
480.00	57.4	56.9	l
500.00	58.5	58.0	l
520.00	59.6	59.4	l
540.00	60.1	60.8	l
560.00	61.7	61.6	l
580.00	63.7	63.0	l
600.00	64.7	64.5	l
620.00	66.1	66.2	l
640.00	66.9	67.5	l
660.00	67.6	68.9	l

Crystalline $C_{sat,m}/R = 0.0960\,T + 5.5_1$ (in temperature range 310 to 464.5 K)
Liquid $C_{sat,m}/R = 0.0635\,T + 26.2_9$ (in temperature range 464.5 to 670 K)

$\Delta^l_c H^o_m(464.5\text{ K}) = 53.3 \pm 2.0$ kJ mol^{-1}

$\Delta^l_c H^o_m(298.15\text{ K}) = 41.7 \pm 3.0$ kJ mol^{-1}

TABLE 10. Continued.

T/K	$C^{II}_{x,m}/R$	T/K	$C^{II}_{x,m}/R$

Naphthalene-2,6-dicarboxylic acid [c]

T/K	$C^{II}_{x,m}/R$	T/K	$C^{II}_{x,m}/R$
315.0	28.9	495.0	42.1
335.0	30.4	515.0	44.2
355.0	32.5	535.0	45.0
375.0	33.9	555.0	46.7
395.0	35.1	575.0	49.2
415.0	36.8	595.0	50.4
435.0	38.0	615.0	51.7
455.0	40.0	635.0	55.0
475.0	41.1	655.0	54.5

Crystalline $C_{sat,m}/R = 0.0763\,T + 4.983$ (in temperature range 305 to 665 K)

Tetraethylsilane

T/K	$C^{II}_{x,m}/R$	$C^{II}_{x,m}/R$	$C^{II}_{x,m}/R$	Phase
mass / g	0.008712	0.016641	0.023425	
V. cell (cm^3)	0.05292	0.05292	0.05288	
315.00	36.8	36.8	36.8	l
335.00	38.2	38.0	38.0	l
355.00	39.7	39.4	39.4	l
375.00	41.4	41.0	40.8	l
395.00	43.2	42.6	42.4	l
415.00	45.2	44.3	44.0	l
435.00	47.3	46.0	45.6	l
455.00	49.5	47.8	47.2	l
475.00	51.6	49.4	48.7	l
495.00	53.7	50.8	49.9	l
515.00	55.8	52.1	50.9	l
535.00	57.7	53.1	51.6	l
555.00	59.6	53.7	51.8	l
575.00	61.4	54.0		l
595.00	63.3			l

[a] With the exceptions of hexan-1,6-diol and tetraethylsilane, all the heat-capacity measurements were made at saturation pressures of less than 0.1 MPa. Therefore, the reported $C^{II}_{x,m}$ values can be assumed to be equal to $C_{sat,m}$ values.

[b] Volume of cell is given for 298.15 K.

[c] Values reported for the crystalline phase only due to decomposition at 695 K (see text).

For tetraethylsilane sample decomposition was greatly reduced by employing a single continuous heat at a heating rate of 0.333 K·s^{-1}, and the abrupt decrease in heat capacity associated with the conversion from the two-phases to one-phase was observed. Temperatures at which conversion to the single phase occurred were measured for five cell fillings. Table 11 reports the density, obtained from the mass of sample and the cell volume calculated with equation 5, and the measured temperatures at which conversion to a single phase was observed. A critical temperature of (606±1) K and a corresponding critical density of (246±5) kg·m^{-3} were derived graphically for tetraethylsilane with these results, as seen in figure 1. Results of measurements on benzene and decane performed as "proof-of-concept measurements" for these procedures have been reported (29). The rapid heating method was used previously for critical temperature and critical density determinations for dibenzothiophene (42).

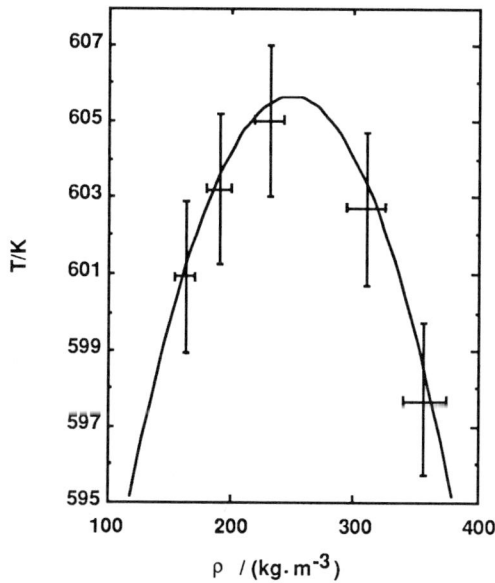

FIGURE 1. Vapor-liquid coexistence region for tetraethylsilane. The crosses span the range of uncertainty.

The critical pressure for tetraethylsilane was not measured directly, but was estimated by means of simultaneous non-linear least-squares fits using the vapor pressures listed in Table 7 and the $C^{II}_{x,m}$ values given in Table 10. $C_{sat,m}$ values were derived using results of the fit and equation (7). Experimental $C^{II}_{x,m}$ were converted to $C^{II}_{v,m}$ values by means of equation (5) for the cell expansion and the vapor-pressure fit described below for $(\delta p/\delta T)_{sat}$. The values of $C^{II}_{v,m}$ were used to derive functions for $(\delta^2 p/\delta T^2)_{sat}$ and $(\delta^2\mu/\delta T^2)_{sat}$. The Cox equation (35) was used to represent the vapor pressures in the form:

$$\ln(p/p_c) = (1 - 1/T_r) \exp(A + BT_r + CT_r^2), \quad (16)$$

with $T_r = T/T_c$, where T_c and p_c are the critical temperature and critical pressure. The critical pressure was included as a variable in the non-linear least-squares analysis. The functional form chosen for variation of the second derivative of the chemical potential with temperature was:

$$(\delta^2\mu/\delta T^2)_{sat} = \sum_{i=0}^{n} b_i(1 - T/T_c)^i. \quad (17)$$

[For compounds where sufficient information was available to evaluate reliably $(\delta^2\mu/\delta T^2)_{sat}$ {e.g., benzene (43), toluene (44)}, four terms (i.e, expansion to n=3) were required to represent the function. Three terms were used in this research.] In these fits the sum of the weighted squares in the following function was minimized:

$$\Delta = C^{II}_{v,m}/R - \{V_m(l)T/nR\}(\delta^2 p/\delta T^2)_{sat}$$
$$+ (T/R)(\delta^2\mu/\delta T^2)_{sat}. \quad (18)$$

For the vapor-pressure fits, the functional forms of the weighting factors used have been reported (20). Within the heat-capacity results, the weighting factors were proportional to the square of the mass of sample used in the measurements. Table 12 lists the coefficients determined in the non-linear least-squares fit. A weighting factor of 20 was used to increase the relative weights of the vapor-pressure measurements in the fit. The weighting factor reflects the higher precision of the vapor-pressure values relative to the experimental heat capacities.

Values of $C_{sat,m}$ for tetraethylsilane were derived from $C^{II}_{v,m}(\rho=\rho_{sat})$ with densities obtained from equation 15 using $\rho_c = 246$ kg·m^{-3}, $T_c = 606$ K, and the acentric factor $\omega = 0.401$. The acentric factor is defined as $\{-\lg(p/p_c) - 1\}$, where p is the vapor pressure at $T_r = 0.7$ and p_c is the critical pressure. The Cox equation coefficients given in Table 8 were used to calculate p. The results for $C_{v,m}^{II}(\rho=\rho_{sat})/R$ and $C_{sat,m}/R$ are reported in Table 13. The estimated uncertainty in these values is 1 per cent.

TABLE 11. Densities and temperatures used to define the vapor/liquid curve near T_c for tetraethylsilane

$\rho/(kg \cdot m^{-3})$	T/K
162.2	600.9
189.7	603.2
230.7	605.0
309.8	602.7
357.2	597.7

TABLE 12. Parameters for equations (16) and (17), critical constants and acentric factor for tetraethylsilane

A	2.50408	b_0	-0.85739
B	-1.28488	b_1	0.87232
C	0.84442	b_2	-2.24261
T_c	606 K	p_c	2400 kPa
ρ_c	246 kg·m^{-3}	ω	0.401

TABLE 13. Values of $C_{v,m}^{II}(\rho = \rho_{sat})/R$ and $C_{sat,m}/R$ for tetraethylsilane ($R = 8.31451$ J·K^{-1}·mol^{-1})

T/K	$C_{v,m}^{II}(\rho = \rho_{sat})/R$	$C_{sat,m}/R$
300.0	35.7	35.7
320.0	36.4	36.4
340.0	37.1	37.1
360.0	37.9	37.9
380.0	38.8	38.8
400.0	39.8	39.8
420.0	41.0	41.0
440.0	42.4	42.4
460.0	44.0	44.1
480.0	46.0	46.2
500.0	48.4	48.6
520.0	51.1	51.6
540.0	54.5	55.3
560.0	58.5	60.0
580.0	63.6	66.9
600.0	70.6	86.0

By judicious choice of starting temperature, the melting endotherms during the d.s.c. enthalpy measurements occurred in the center of a heating cycle. The measured enthalpies during those particular heating cycles contained the enthalpy of fusion plus enthalpies for raising the solid from the initial temperature to the melting point and for raising the liquid from the melting point to the final temperature. Figure 2 shows the thermochemical cycle used to "correct" the enthalpy of fusion of the compound at its melting point to the corresponding value at the standard temperature of 298.15 K. Details of the derived enthalpies of fusion for tetradecan-1-ol, hexan-1,6-diol, benzoyl formic acid, and naphthalene-2,6-dicarboxylic acid dimethyl ester at their melting points and the corresponding values at 298.15 K are reported in Table 10. Equations (representing the heat capacities for both the liquid and solid phases for each compound) which were used in the "correction" to 298.15 K are also reported in Table 10. [Note the heat-capacity equations should only be used to derive values within the temperature ranges specified in Table 10; extrapolation outside the temperature range will produce erroneous values. As an extreme example, extrapolation of the reported solid-phase heat capacity equation for tetradecan-1-ol by 100 K gives a negative heat capacity.]

The sample of methacrylamide polymerized at temperatures greater than 392 K. At 390 K its vapor pressure was 2.38 kPa precluding vapor-pressure measurements in the ebulliometer (pressure range 2 to 270 kPa). Inclined-piston vapor-pressure measurements made in the temperature range 325 K to 390 K are reported in Table 14. Subsequent d.s.c. enthalpy measurements, also reported in Table 14, defined the melting point of the sample as 385.1 K. Hence, the two vapor-pressure measurements at 385 K (due to some premelting) and 390 K, were for the liquid phase. Since the sublimation pressure measurements extended only over a 50 K range it was inappropriate to fit any vapor-pressure equation other than the "simple" Clausius Clapeyron;

$$\ln(p/1\text{kPa}) = A + B/(T/K). \qquad (19)$$

Results of the fit are given in Table 14. Values of "A" and "B" in Equation 19 were determined also for the liquid phase with the two experimental values that were obtained before polymerization. The corresponding enthalpies of sublimation and vaporization (calculated from the values of B obtained) at the mid-temperature of the data range are also reported in Table 14. The intersection point of the solid- and liquid-phase vapor-pressure curves was 385.1 K, in excellent agreement with the d.s.c. value for the fusion temperature.

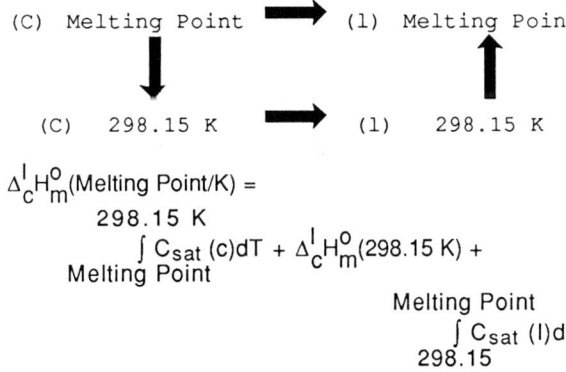

FIGURE 2. Thermochemical cycle relating $\Delta_c^l H_m^o(298.15\ K)$ to $\Delta_c^l H_m^o(\text{melting point K})$.

TABLE 14. Vapor-pressure and heat-capacity measurements, derived enthalpies of sublimation, vaporization and fusion for methacrylamide. "IP" refers to measurements made with the inclined piston, Δp is the difference of the calculated value of pressure {ln(p/1kPa) versus 1/(T/K) equation fit} from the observed value of pressure ($p - p_{Fit}$), $\sigma(p)$ is the propagated error calculated from Equation 3.

Material	$\dfrac{T}{K}$	$\dfrac{p}{kPa}$	$\dfrac{\Delta p}{kPa}$	$\dfrac{\sigma(p)}{kPa}$
IP	325.000	0.0137	0.0001	0.0001
IP	335.000	0.0347	-0.0001	0.0001
IP	345.000	0.0831	-0.0010	0.0001
IP	355.007	0.1929	-0.0009	0.0001
IP	365.000	0.4270	0.0009	0.0002
IP	375.000	0.9030	-0.0042	0.0002
	Triple point	**385.1 K**		
IP	385.003	1.8201		0.0005
IP	390.001	2.3803		0.0006

$$\ln(p/1\text{kPa}) = 27.136 - 10216/(T/K)$$
$$\text{crystalline} \rightarrow \text{gas phase}$$
$$\ln(p/1\text{kPa}) = 21.54 - 8061/(T/K)$$
$$\text{liquid} \rightarrow \text{gas phase}$$

$$\Delta_c^g H_m^o(350\ K) = 84.9 \pm 0.4\ \text{kJ mol}^{-1}$$
$$\Delta_l^g H_m^o(387.5\ K) = 67.0 \pm 0.4\ \text{kJ mol}^{-1}$$

T/K	$C_{x,m}^{II}/R$	$C_{x,m}^{ii}/R$	Phase
mass / g	0.023179	0.010071	
V. cell (cm³)	0.05288	0.05292	
295.0	17.2	17.3	cr
305.0	17.6	17.5	cr
315.0	18.0	17.9	cr
325.0	18.3	18.4	cr
335.0	18.6	18.6	cr
345.0	19.0	19.0	cr
355.0	19.3	19.3	cr
365.0	19.8	19.7	cr

Crystalline phase
$C_{sat,m}/R = 6.82 + 0.0352\ T$
(in temperature range 290 K to 385 K)

$$\Delta_c^l H_m^o(385.1\ K) = 15.0 \pm 1.0\ \text{kJ mol}^{-1}$$

DISCUSSION

Table 15 summarizes the thermochemical property measurements and derived ideal-gas standard enthalpies of formation for all the compounds of this study. In this section of the report the results obtained for each compound are discussed and compared with previously available literature values and relevant group-contribution parameters derived.

TABLE 15. Thermochemical properties at 298.15 K. Values are in kJ·mol^{-1}.
(R = 8.31451 J·K^{-1}·mol^{-1} and p° = 101.325 kPa) [a]

	$\Delta_f H^o_m$(c)	$\Delta_c H^l_m$	$\Delta_f H^o_m$(l)
A			-342.6±0.3
B	-628.18±1.46	48.5±1.0	-579.7±1.8
C	-583.86±0.72	21.6±1.0	-562.3±1.2
D	-247.10±0.50		
E	-487.22±0.80	19.4±0.6	-467.8±1.0
F	-770.25±0.96	?.?	[c]
G	-696.51±1.34	41.7±3.0	-654.8±3.3
H			-328.6±5.2

	$\Delta_f H^o_m$(l)	$\Delta_l^g H_m$	$\Delta_f H^o_m$(g)
A	-342.6±0.3	49.86±0.03	-292.7±0.3
B	-579.7±1.8	104.9±1.9	-474.8±2.6
C	-562.3±1.2	102.9±1.5	-459.4±1.9
D		89.4±2.0[b]	-157.7±2.1
E	-467.8±1.0	79.12±0.08	-388.7±1.0
F		?.?	?.?
G	-654.8±3.3	99.67±5.15	-555.1±6.1
H	-328.6±5.2	44.62±0.05	-284.0±5.2

[a] A = (±)-butan-2-ol; B = tetradecan-1-ol; C = hexan-1,6-diol: E = benzoyl formic acid F = naphthalene-2,6-dicarboxylic acid G = naphthalene-2,6-dicarboxylic acid dimethyl ester and H = tetraethylsilane.
[b] Value for enthalpy of sublimation at 298.15 K; see text and Table 17.
[c] Compound decomposes before the melting point; see text.

Butan-2-ol

Butan-2-ol is optically active. The results reported here are for the racemic (R,S) mixture. (±)-Butan-2-ol has been the subject of previous combustion calorimetric (45), vapor-pressure (40,46-49), and heat-capacity studies (50,51). The major reason for the inclusion of the compound in this study was to confirm the previous measurements and solidify our knowledge of the value of the $-CH_2-$ group increment in the alcohol series. Sunner and coworkers (52,53) in papers published in the late 1970's drew attention to the "abnormally" low value for the CH_2 group increment in the alcohols. At that time sufficient uncertainty existed, particularly in the enthalpies of vaporization at 298.15 K, to cloud the picture somewhat.

The energy of combustion for (±)-butan-2-ol reported in this research -35827.4±1.2 J·g^{-1} is in excellent agreement with that obtained by Skinner and Snelson (45) -35828.5±6.2 J·g^{-1} in their study on the four isomeric butyl alcohols. Vapor-pressure measurements made on (±)-butan-2-ol by the research group at the National Physical Laboratory (NPL) at Teddington, Middlesex, England have been reported four times in the literature (40,46-48). Biddiscombe et al. (46) reported values obtained in the temperature range 345 K to 380 K; Ambrose and Townsend (40) reported measurements in the temperature range 422 K to 535.95 K (from 474 kPa to the critical pressure 4194 kPa); Ambrose and Sprake (47) reported measurements made on (±)-butan-2-ol (341 K to 380 K), one of a series of selected alcohols from methanol through hexadecan-1-ol; and the same authors (48) also reported results from an ebulliometric study (307 K to 381 K) on the pure optical isomer (+)-butan-2-ol. Figure 3 compares the above vapor-pressure measurements (except those of Ambrose and Townsend (40) which fall outside the range of the measurements reported here) with those obtained in this research. On average the NPL results are slightly lower than those reported here but the agreement remains excellent (±0.15 percent). The results also confirm the statement of Ambrose and Sprake (48) that the vapor pressure of pure "(+)-butan-2-ol is not significantly different from that of (±)-butan-2-ol." Vapor pressure measurements reported by Brown, Foch, and Smith (49) (323 K to 373 K, 10.6 kPa to 101.3 kPa) are consistently 0.4 percent greater than values calculated using the Cox equation coefficients given in Table 8.

The derivation of reliable values for the enthalpy of vaporization of (±)-butan-2-ol at temperatures within the range of the vapor-pressure measurements requires accurate values for the difference

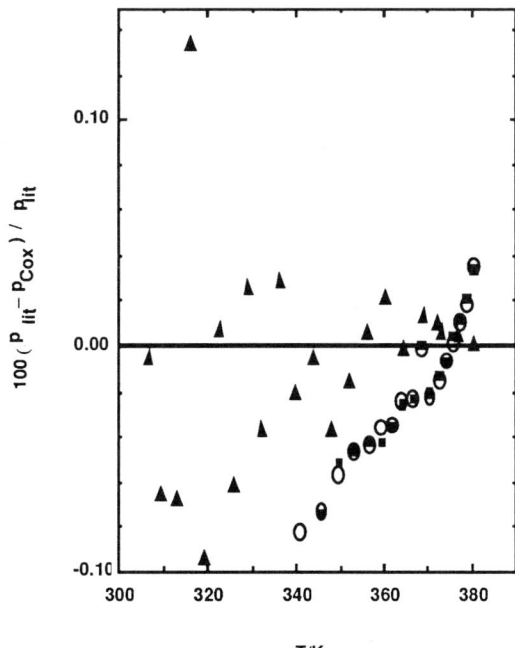

FIGURE 3. Comparison of measured vapor pressures for (±)-butan-2-ol with those obtained previously at NPL. P_{lit} is the literature value of the vapor pressure. ■ Biddiscombe et al. (46); ○ Ambrose and Sprake (47); ▲ Ambrose and Sprake (48).

between the molar volumes of the real gas and the liquid, ($\Delta_l^g V_m$ in equation 14). The method used to calculate values of $\Delta_l^g V_m$ in this research was outlined above and depends on extended corresponding states being applicable over the temperature range under consideration. The estimated second viral coefficients obtained from extended corresponding states were in good agreement (±8 percent) with those measured by Cox (54). Densities for the liquid obtained using equation 15 and the critical constants given in Table 9A are compared with literature values (55-57) in Figure 4. Above 350 K the values reported by Costello and Bowden (55) differ by over 3 percent with those reported by Hales and Ellender (56). The latter agree with those calculated in this research using extended corresponding states to within a few tenths of a percent. Below 350 K, the deviation of extended corresponding states from the experimentally measured densities may be due to the effects of hydrogen bonding. The calculated enthalpies of vaporization of this research are not affected significantly by the uncertainty in the liquid-phase densities.

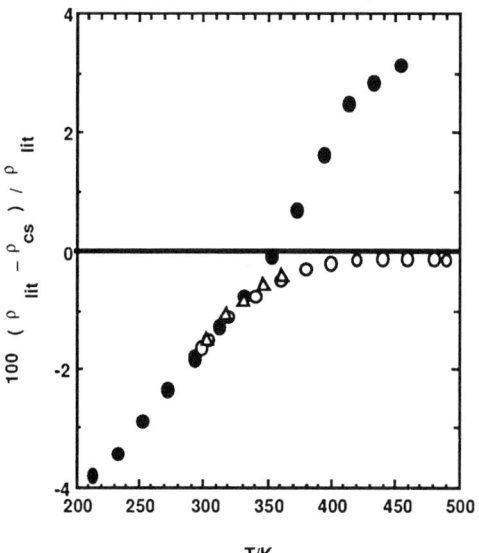

FIGURE 4. Comparison of literature density measurements for (±)-butan-2-ol with those obtained using extended corresponding states. ρ_{cs} = value of the density calculated using equation 15 and the critical properties reported in Table 9A. ● Costello and Bowden; (55); ○ Hales and Ellender (56); △ Thomas and Meatyard (57).

Figure 5 compares the enthalpies of vaporization derived in this research (Table 9) with values determined experimentally by vaporization calorimetry (51,58,59). With the exception of the single datum point of McCurdy and Laidler (59) the agreement is good, especially above 320 K.

A full discussion of the group parameters used to estimate the ideal-gas enthalpy of formation of (±)-butan-2-ol is given later in this report in the discussion on tetradecan-1-ol. It is, however, worth noting at this stage that the $-CH_2-$ group increment obtained by differencing the enthalpy of formation reported in Table 15 and that for propan-2-ol [as assessed by Pedley, Naylor, and Kirby (60)] is -19.9 ± 0.6 kJ·mol^{-1} compared to the "universal" value of -20.72 kJ·mol^{-1} given by the group-additivity parameters listed in references 1 and 2.

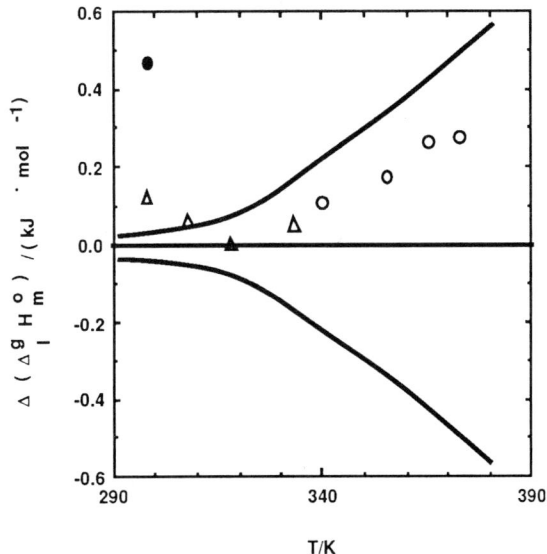

FIGURE 5. Comparison of experimentally measured enthalpies of vaporization for (±)-butan-2-ol with those calculated using the Clapeyron equation. $\Delta(\Delta_l^g H_m^o) = \{\Delta_l^g H_m^o(\text{literature}) - \Delta_l^g H_m^o(\text{this research})\}$. The curves represent the uncertainty limits (approximately two standard deviations) on the enthalpies of vaporization reported in this research. ○ Berkman and McKetta; (51); △ Polák and Benson (58); ● McCurdy and Laidler (59).

Tetradecan-1-ol

The literature on polymorphism in the higher alkan-1-ols in general and tetradecan-1-ol in particular is large. References 61 through 68 are some examples of studies in this field. Contradictions still exist on the relative stability of the crystal phase present at a given temperature (67,68). The sample of tetradecan-1-ol used in this study exhibited the same properties as that designated as the β form by Mosselman et al. (69).

Mosselman and Decker (70) reported energy of combustion measurements on the C_{12} through the C_{16} alkan-1-ols in which the energy of combustion of tetradecan-1-ol is listed as -9149.6 ± 1.2 kJ·mol^{-1} for the β form. This is in good agreement with the value -9151.06 ± 1.16 kJ·mol^{-1} obtained in this research. Mosselman and Decker (70) stated; "Reduction to the vapor phase for the normal primary alkanols as a group is not yet possible, since reliable enthalpies of vaporization are known only for compounds up to and including C_6."

Vapor-pressure measurements on tetradecan-1-ol both above and below its triple point have been reported in the literature by several research groups (71-74). Further discussion of the solid-phase measurements is given later in this report when the enthalpy of sublimation is discussed. The liquid-phase literature data fall into two groups: 1) measurements made close to the triple point (311 K to 350 K); and 2) measurements made at higher temperatures (pressures) in the region 450 K to approximately 600 K. Figures 6 and 7 compare both groups with values obtained using the Cox equation coefficients listed in Table 8.

For the data below 360 K (figure 6) the measured vapor pressures are scattered and, with the exception of one point, are significantly higher than values obtained using the Cox equation coefficients. It must be noted, however, that this represents an appreciable (~100 K) extrapolation and the vapor pressures in this region are small (of the order 0.07 to 4.6 Pa). For the data above 400 K, as shown in Figure 7, with the exception of two points [424.75 K/0.69 kPa (18.3 percent low) and 422.3 K/6.45 kPa (3.6 percent high) not plotted in the figure], the vapor-pressure measurements of Kemme and Kreps (73) are in good agreement with values calculated using the Cox equation coefficients. Ambrose and Walton (74) list values for the critical temperatures, critical pressures, and coefficients for the Wagner equation (75) for the vapor pressures of both the normal alkanes

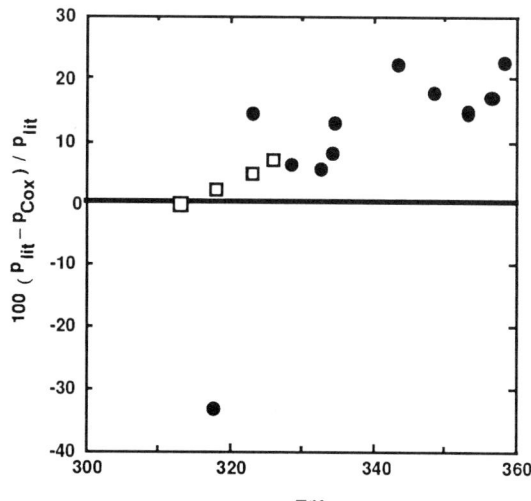

FIGURE 6. Vapor-pressure comparisons for tetradecan-1-ol in the "low" pressure region. Literature vapor-pressure values are compared with those obtained using the Cox equation coefficients listed in Table 8.
□; Davies and Kybett (71)
●; Spizzichino (72).

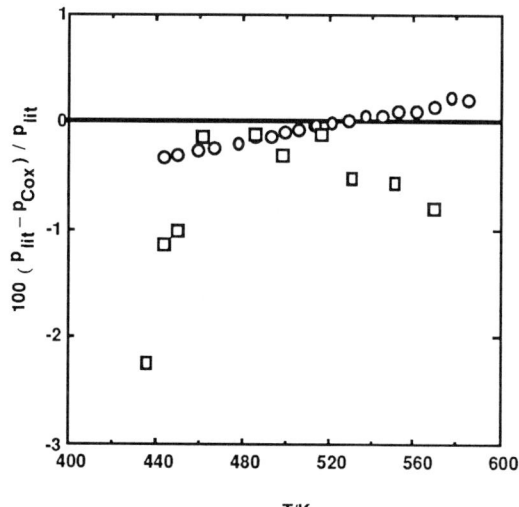

FIGURE 7. Vapor-pressure comparisons for tetradecan-1-ol in the "high" pressure region. Literature vapor-pressure values are compared with those obtained using the Cox equation coefficients listed in Table 8.
□; Kemme and Kreps (73)
○; Ambrose and Walton (74).

and the alkan-1-ols. The listed coefficients for tetradecan-1-ol were obtained by linear interpolation (see reference 75). The Wagner equation and the coefficients listed by Ambrose and Walton were used to calculate vapor pressures at the temperatures of the measurements reported in this research. The percentage deviation between the calculated and observed vapor pressures is shown in figure 7. The maximum deviation is only 0.3 percent for the values derived using the Ambrose and Walton equation! This would seem to attest to the accuracy of the representation of the vapor pressures of the alkan-1-ols derived by Ambrose and Walton.

As noted earlier, the derivation of reliable values for the enthalpy of vaporization within the range of the vapor-pressure measurements requires accurate values for the difference between the molar volumes of the real gas and the liquid, $\Delta_l^g V_m$. The method used to calculate values of $\Delta_l^g V_m$ in this research was outlined above and depends on extended corresponding states being applicable over the temperature range under consideration. Densities for the liquid obtained using equation **15** with the critical constants given in Table 9A are compared with literature values (55,76) in Figure 8. The two literature datasets are in good agreement with each other but those of Matsao and Makita (76) cover only a short temperature range. Over the temperature range 313 K to 573 K the percentage deviations between the observed [Costello and Bowden (55)] and the extended corresponding states values fall on a monotonic curve ranging from +1 percent to -1 percent. Derived enthalpies of vaporization for tetradecan-1-ol are listed in Table 9.

Månsson et al. (77) measured the enthalpy of vaporization at temperatures just above the melting point and, using heat-capacity data from Mosselman et al. (69), corrected the experimental measurements to 298.15 K. They reported a value $(\Delta_l^g H_m^o \: C_{14}H_{30}O, \: 298.15 \: K) = 102.2 \pm 2.3$

kJ·mol^{-1} which compares favorably with the value obtained in this research 104.9±1.9 kJ·mol^{-1}. Spizzichino (72) derived an equation to represent the enthalpy of vaporization over the temperature range 313 K to 373 K:

$$\Delta_l^g H_m^o /(kJ\cdot mol^{-1}) = 174.81 - 0.234\, T . \quad (20)$$

Davis and Kybett (71) derived from their vapor-pressure measurements a value $(\Delta_l^g H_m^o\, C_{14}H_{30}O,\, 320\, K) = 104.2 \pm 1.7$ kJ·mol^{-1} and Kramer (78) reported $(\Delta_l^g H_m^o\, C_{14}H_{30}O,\, 566.2\, K) = 66.92$ kJ·mol^{-1}. Figure 9 compares the literature values with those derived in this research. No major deviations are apparent except for the value quoted by Kramer.

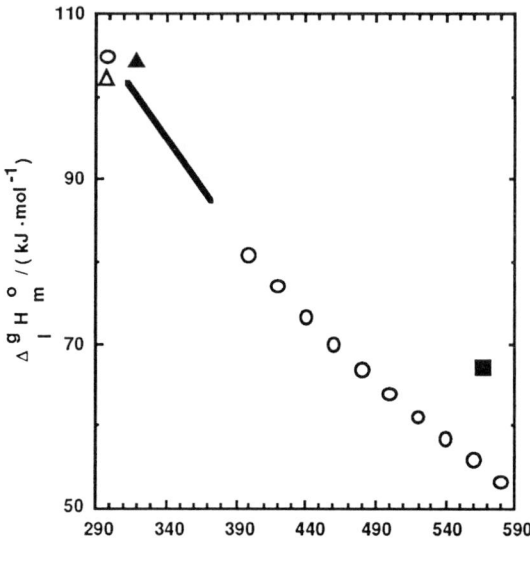

FIGURE 9. Comparison of literature enthalpy of vaporization measurements for tetradecan-1-ol with those obtained in this research. O; This research, $\triangle$; Månnson et al. (77), ▲; Davis and Kybett (71), ■; Kramer (78), the solid line; Spizzichino (72).

Since the vapor pressure was below 0.1 MPa over the range of the $C_{x,m}^{II}$ measurements reported in Table 10 no correction for vaporization of the sample into the vapor-space of the d.s.c. cells was made. The saturation heat capacities $C_{sat,m}$ were assumed to be the mean of the reported values. Figure 10 compares the $C_{sat,m}$ values with published values (69,79,80). Differences are within the relative uncertainties in the measurements (81).

The enthalpy of fusion, $(\Delta_c^l H_m^o\, C_{14}H_{30}O,\, 311\, K) = 49.4\pm0.4$ kJ·mol^{-1}, listed in Table 10 is in good agreement with the value of 49.51±0.42 kJ·mol^{-1} obtained by Mosselman et al. (69) which lends credence to the assignment of the β form to the crystalline state of the sample. Combination of the listed enthalpies of fusion (Table 10) and vaporization at 298.15 K (Table 15) gives a value $(\Delta_c^g H_m^o\, C_{14}H_{30}O,$

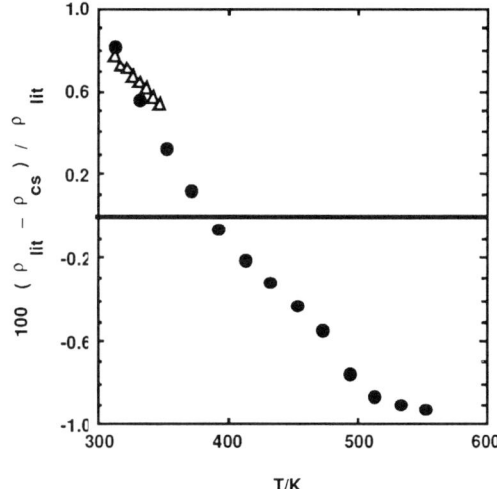

FIGURE 8. Comparison of literature density measurements for tetradecan-1-ol with those obtained by extended corresponding states. ρ_{cs} = value of the density calculated using equation 15 with the critical properties reported in Table 9A. ●; Costello and Bowden; (55); $\triangle$; Matsao and Makita (76).

298.15 K) = 153.4±2.1 kJ·mol^{-1}. With solid-phase vapor-pressure measurements Davis and Kyber (71) obtained a value ($\Delta_c^g H_m^o$ C$_{14}$H$_{30}$O, 300 K) = 143.9 ±2.1 kJ·mol^{-1}. Hoyer and Peperle (82) using the same procedure, but with a different set of vapor-pressure measurements, reported ($\Delta_c^g H_m^o$ C$_{14}$H$_{30}$O, 310.7 K) = 159.4 kJ·mol^{-1}. The deviations attest to the difficulty in making accurate vapor-pressure measurements below 10 Pa.

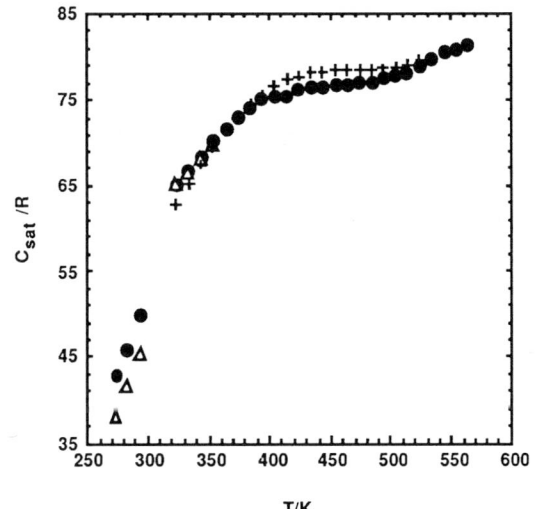

FIGURE 10. Comparison of literature heat-capacity measurements for tetradecan-1-ol with those obtained in this research. ●; This research, △; Mosselman, Mourik, and Dekker (69), +; Vasil'ev et al (79) which are the same as Khasanshin et al. (80).

Table 16 lists ideal-gas enthalpies of formation for the alkan-1-ols from ethanol through hexadecan-1-ol. Column 2 of Table 16 lists enthalpies of formation reported by NPL (83) and, in the case of tetradecan-1-ol, this research. Column 3 lists the "Assessed Best" values reported by Pedley, Naylor, and Kirby (60) and column 4 lists values estimated using the updated group-additivity parameters of Benson listed in reference 2. The difference between column 4 and either column 2 or 3 increases monotonically as the size of the alkan-1-ol increases. Column 5 lists the values obtained using the equation;

$$[\Delta_f H_m^o, CH_3(CH_2)_n OH, g\ 298.15\ K]/(kJ \cdot mol^{-1}) = -214.57 - 20.206\,n \quad (21)$$

for n≥1. The agreement obtained is excellent. The –CH$_2$– increment {the [C-(C)$_2$(H)$_2$] group-additivity parameter} is then 0.51 kJ·mol^{-1} smaller than the "universal" value of –20.72 kJ·mol^{-1} listed by Benson (1,2). The –CH$_2$– group increment obtained above for alkan-2-ols, –19.9±0.6 kJ·mol^{-1}, is in excellent agreement with this lower value.

TABLE 16. Enthalpies of formation of alkan-1-ols. Comparison of experimental (Expt.),[a] assessed (A),[b] and estimated by Benson groups (BG)[c] ideal-gas enthalpies of formation at 298.15 K for some alcohols. All values in kJ·mol^{-1}.

Compound	Expt	A	BG	Eq21
Ethanol		–235.2±0.4	–234.8	–234.8
Propan-1-ol		–255.1±0.5	–255.5	–255.0
Butan-1-ol	–275.28±0.53	–275.0±0.4	–276.2	–275.2
Pentan-1-ol	–295.63±0.74	–294.5±0.5	–297.0	–295.5
Hexan-1-ol		–315.8±0.6	–317.7	–315.6
Heptan-1-ol		–336.4±1.0	–338.4	–335.8
Octan-1-ol	–356.87±1.17	–355.5±0.8	–359.1	–356.0
Nonan-1-ol		–376.3±1.4	–379.8	–376.2
Decan-1-ol		–396.4±1.6	–400.6	–396.4
Dodecan-1-ol		–436.6±1.1	–442.0	–436.8
Tetradecan-1-ol	–474.8±2.6	–478.4±2.4	–483.4	–477.2
Hexadecan-1-ol	–517.5±3.2	–517.0±2.4	–524.9	–517.7

[a] Reference 83 except for tetradecan-1-ol the value for which was obtained in this research.
[b] Reference 60.
[c] References 1 and 2.

Hexan-1,6-diol

When this research was started, one thermochemical study (84) on hexan-1,6-diol existed in the literature. The reported vapor-pressure results are inconsistent with the equation given failing to reproduce the listed normal boiling point. The "correction" of the derived enthalpy of vaporization at the mid temperature of the vapor-pressure measurements to give a value at 298.15 K also appear inconsistent.

Since this research was completed, an energy of combustion, and enthalpies of fusion and sublimation measurements on hexan-1,6-diol have been reported by Knauth and Sabbah (85, 86).

The energy of combustion, $\{(\Delta_c U_m^o/M)\}/(J \cdot g^{-1}) = -31916.7 \pm 1.6$, obtained in this research is lower than either of the two reported values, -32087 ± 41 $J \cdot g^{-1}$ and -31999 ± 37 $J \cdot g^{-1}$ of references 84 and 85, respectively. In neither study was the sample purity sufficient for accurate measurements. Gardner and Hussain (84) list a sample purity of 99.8 mole percent and Knauth and Sabbath (85) state a purity of only 99.38±0.02 mole percent. In contrast glc analysis of the sample used in this research gave a purity of >99.95 percent which was corroborated both by the CO_2 analyses (Table 2) during the combustion measurements and the small differences observed between the boiling and condensation temperatures (Table 7) during the ebulliometric vapor-pressure measurements. The spread of measured energies of combustion observed by Knauth and Sabbah was -32116 $J \cdot g^{-1}$ to -31896 $J \cdot g^{-1}$, which encompasses all of the values obtained here where the spread (Table 5) was -31924 $J \cdot g^{-1}$ to -31911 $J \cdot g^{-1}$. Gardner and Hussain do not give sufficient details to delineate their spread in values, but from the reported uncertainty interval, the width must have been similar to that of Knauth and Sabbah.

The vapor-pressure equation given by Gardner and Hussain cannot be reconciled with their listed normal boiling point. Consistency is obtained if it is assumed that their constant "B" contains a typographical error and should read 9.37±0.18 instead of 8.37±0.18. Then the equation would read:

$$\lg(p/\text{torr}) = 9.37 - 3405/(T/K)$$
(in the temperature range 451 K to 525 K), (22)

where 1 torr = (101.325/760) kPa. Figure 11 compares the vapor pressures calculated using equation **22** with the values reported in Table 7. Gardner and Hussain assigned an uncertainty of ±2 percent on their measurements. Above 480 K the agreement is within their assigned uncertainty limits.

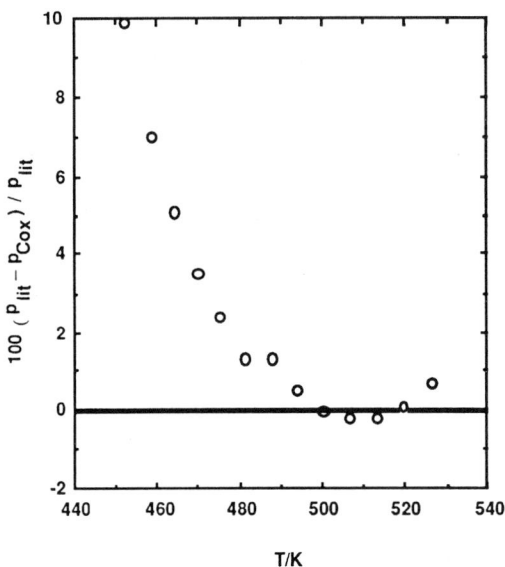

FIGURE 11. Comparison of hexan-1,6-diol vapor-pressure measurements with literature values. The "literature values" are those calculated using the "corrected" (see text and equation **22**) vapor-pressure equation of Gardner and Hussain (84).

The enthalpy of vaporization at the mid temperature of the vapor-pressure measurements listed by Gardner and Hussain ($\Delta_l^g H_m^o$ $C_6H_{14}O_2$, 488 K) = 65.2±1.2 $kJ \cdot mol^{-1}$ is in excellent agreement with an interpolated value (see Table 9) of ($\Delta_l^g H_m^o$ $C_6H_{14}O_2$, 488 K) = 65.6±0.3 $kJ \cdot mol^{-1}$. Gardner and Hussain list a value ($\Delta_l^g H_m^o$ $C_6H_{14}O_2$, 298.15 K) = 83.3±1.7 $kJ \cdot mol^{-1}$ obtained using the Watson equation (87) to convert the 488 K result to 298.15 K. This result is in stark contrast to the value obtained in this research ($\Delta_l^g H_m^o$ $C_6H_{14}O_2$, 298.15 K) = 102.9±1.5 $kJ \cdot mol^{-1}$ (Table 9).

The measured enthalpy of fusion 22.6±0.6 $kJ \cdot mol^{-1}$ of this research (Table 10) is in good agreement with the value reported by Knauth and Sabbah, 22.2±0.3 $kJ \cdot mol^{-1}$ and somewhat lower than that of Gardner and Hussain 25.5±0.3 kJ·mol^{-1}. Knauth and Sabbath measured enthalpies of sublimation and list ($\Delta_c^g H_m^o$ 298.15 K) = 112.0±0.4 kJ·mol^{-1} which is somewhat lower than the value ($\Delta_c^g H_m^o$ C$_6$H$_{14}$O$_2$, 298.15 K) = 124.5±1.8 kJ·mol^{-1} obtained by addition of the enthalpies of vaporization and fusion reported here in Tables 9, 10, and 15. These differences may be due to the lower purity of the sample used by Knauth and Sabbah.

Addition of the group-additivity parameters (1,2) for hexan-1,6-diol follows:

2	C-(C)(O)(H)$_2$	-33.91 times 2	-67.82
2	O-(C)(H)	-158.68 times 2	-317.36
4	C-(C)$_2$(H)$_2$	-20.21 times 4	-80.84
($\Delta_f H_m^o$ C$_6$H$_{14}$O$_2$, g 298.15 K)			-466.0 kJ·mol^{-1}

using the C(C)$_2$(H)$_2$ group increment derived above for the alkanols. The value (Table 15) obtained in this research is ($\Delta_f H_m^o$ C$_6$H$_{14}$O$_2$, 298.15 K) = -459.4±1.9 kJ·mol^{-1}, which is 6.6 kJ·mol^{-1} more positive than that estimated using the group-additivity parameters. The difference is larger than normal (±4 kJ·mol^{-1}) and could be due to any one or a combination of three effects: 1) misassigned parameter(s), 2) a destabilizing effect within the molecule which is not allowed for in the estimation, or 3) experimental error. Changing the group parameters to fit the experimental result would have a "ripple effect" making the estimation of the alkanols (Table 16) diverge from the present good agreement. Byström and Månsson (88) have noted destabilization effects in other oxygen-containing compounds and intramolecular hydrogen bonding may be present in this case. However, that is pure speculation in the absence of a structural analysis in the gas phase. Experimental error is always possible and may in this case arise in the long extrapolation necessary for the derivation of the enthalpy of vaporization at 298.15 K. However, for tetradecan-1-ol, where a similar extrapolation was necessary,

the agreement with literature values was good (loc. cit).

Methacrylamide

Results of a previous study of the thermochemical properties of methacrylamide have been reported in the literature (89). Lebedeva, Gutner, and Kiseleva (89) gave little detail of the measurements made and list an energy of combustion of 27284±6 J·g^{-1} compared to the value 27324.6±2.1 J·g^{-1} (Table 5) obtained in this research. Lebedeva et al. list an enthalpy of fusion of 17.6±0.8 kJ·mol^{-1}, which is considerably higher than that obtained in this research ($\Delta_c^l H_m^o$ C$_4$H$_7$NO, 385.1 K) = 15.0±1.0 kJ·mol^{-1} (Table 14). Polymerization of our sample started at 392 K so the difference may be due to a portion of the enthalpy of polymerization being contained within the fusion measurements made by Lebedeva et al., (89).

Table 17 outlines "correction" of the measured enthalpy of sublimation at 350 K listed in Table 14 to both 385.1 K (the melting point) and 298.15 K using a cycle similar to that shown in Figure 2. In the absence of any experimental measurements the ideal-gas heat capacity for methacrylamide used in Table 17 was obtained by taking the values listed by Stull, Westrum and Sinke (90) for acrylic acid and multiplying them by the ratio of the respective molar masses.

The small difference (0.4 kJ·mol^{-1}) between the measured enthalpy of fusion and the value calculated in the thermochemical cycles in Table 17 illustrates the self-consistency of the measurements. This, in turn, gives confidence to the derived enthalpy of sublimation at 298.15 K given in Tables 15 and 17.

The ideal-gas enthalpy of formation at 298.15 K is:

($\Delta_f H_m^o$ C$_4$H$_7$NO, g 298.15 K) = -157.7±2.1 kJ·mol^{-1}.

Addition of the group-additivity parameters (1,2,91) follows;

N–(CO)(H)$_2$ –67.82
CO–(C$_d$)(N)   –115.3
C$_d$–(C$_d$)(C)(CO)   39.36
C–(C$_d$)(H)$_3$ –42.20
C$_d$–(C$_d$)(H)$_2$ 26.21

gives ($\Delta_f H_m^o$ C$_4$H$_7$NO, g 298.15 K)

                                          –154.3 kJ·mol^{-1},

using the CO–(C$_d$)(N) value derived in 1987 in this research program from the results for acrylamide. In 1987, since the compound polymerized so easily on heating above room temperature, some doubt existed in the accuracy of the literature vapor pressures (92) used to derive the ideal-gas enthalpy of formation of acrylamide. The difference between the experimental and group-additivity estimate, 3.4 kJ·mol^{-1}, is within the uncertainty intervals assigned to the measurements. Therefore, the doubt in the accuracy of the literature vapor pressures for acrylamide is erased. Combining the acrylamide and methacrylamide results, an average value of –117±2 kJ·mol^{-1} is recommended for the CO–(C$_d$)(N) group.

Table 17. Thermochemical cycles for methacrylamide

Sublimation at 298.15 K

$\Delta_c^g H_m^o$(350 K) =

$$\int_{350}^{298.15} C_{sat}(c)dT + \Delta_c^g H_m^o(298.15\,K)$$
$$+ \int_{298.15}^{350} C_{sat}(g)dT$$

$C_{sat}/R = 6.82 + 0.0352\,T$ for crystalline phase
(see Table 14)
$C_{sat}/R = 4.13 + 0.0234\,T$ for the gas phase (see text)
$\Delta_c^g H_m^o(350\,K) = 84.9 \pm 0.4$ kJ·mol^{-1} (see Table 14)

Hence:
    84.9 = –12.8 + $\Delta_c^g H_m^o$(298.15 K) + 8.3   in kJ·mol^{-1}
and;
    $\Delta_c^g H_m^o$(298.15 K) = 89.4 ± 2.0 kJ·mol^{-1}

Table 17. Continued

Sublimation at 385.1 K

$\Delta_c^g H_m^o$(350 K) =

$$\int_{350}^{385.1} C_{sat}(c)dT + \Delta_c^g H_m^o(385.1\,K) + \int_{385.1}^{350} C_{sat}(g)dT$$

Hence:
    84.9 = 9.5 + $\Delta_c^g H_m^o$(385.1 K) – 6.2   in kJ·mol^{-1}
and;
    $\Delta_c^g H_m^o$(385.1 K) = 81.6 ± 2.0 kJ·mol^{-1}

which compares favorably with sum of the enthalpies of vaporization and fusion reported in Table 14
    (67.0 and 15.0 kJ·mol^{-1}).

Naphthalene-2,6-dicarboxylic acid dimethyl ester

The discussion of naphthalene-2,6-dicarboxylic acid dimethyl ester follows out of the compound order used so far in this report to set the logic for some modifications of group parameters given in references 1 and 2 before discussing benzoyl formic acid. Throughout this section naphthalene-2,6-dicarboxylic acid dimethyl ester will be denoted as NDCA-2M.

In the literature, reports of the physical and thermodynamic properties of NDCA-2M are rare. Freund and Fleischer (93) in 1913 reported a melting point of 461 K for this compound. More recently (1978) Dozen, Fujishoma, and Shingu (94) measured values of 460-1 K (in a capillary tube) and 464.4 K in a d.s.c. The latter value is in excellent agreement with that observed in this research, 464.5 K (Table 10). Dozen et al. also measured an enthalpy of fusion, 38.4 kJ·mol^{-1}, but state that the sample lost 7.7 percent of its original mass during the measurement. The enthalpy of fusion measured in this research, 53.3 ± 2.0 kJ·mol^{-1}, is 39 percent higher than that reported by Dozen et al. (94).

The thermochemical properties of DNCA-2M at 298.15 K are listed in Table 15. When the group-additivity parameters (1,2) were initially summed for this compound, the estimated ideal-gas enthalpy of formation at 298.15 K obtained, -523.3 kJ·mol^{-1}, was substantially different from that given in Table 15; -555.1 ± 6.1 kJ·mol^{-1}. Further investigation of the various group parameters involved followed.

First, it was noted that the difference between the "assessed" (60) ideal-gas enthalpies of formation for the aliphatic carboxylic acids and the corresponding methyl esters averaged 21.3 ± 2.7 kJ·mol^{-1} (Table 18). The group-additivity parameters listed in references 1 and 2 gave 15.48 kJ·mol^{-1} for the difference:

O–(CO)(H)   O–(CO)(C) + C(O)(H)$_3$
 -243.25 -185.48 + -42.29 kJ·mol^{-1}

It was assumed that the "error" lay in the O–(CO)(C) group parameter. It was assigned a value of -179.7 kJ·mol^{-1} i.e., $-185.48 + (21.3-15.48)$ (previously -185.48 kJ·mol^{-1}). Table 18 lists the enthalpies of formation of the carboxylic acids, their methyl esters, and values for the difference between the respective enthalpies of formation.

Next it was noted that the group parameters listed in references 1 and 2 failed to reproduce the ideal-gas enthalpies of formation at 298.15 K for either benzoic acid (60) or naphthalene-2-carboxylic acid (95): c.f., -294.1 ± 2.2 (60) with -269.6 and -223.1 ± 1.0 (95) with -201.8, respectively (all values in kJ·mol^{-1}). The relevant group parameters were:

C_b–$(C_b)_2$(CO) + CO–(C_b)(O)
  40.61 -136.07 kJ·mol^{-1}.

It was assumed that the "error" lay in the C_b–$(C_b)_2$(CO) group parameter. It was assigned a value of 17.5 kJ·mol^{-1} rather than 40.61 kJ·mol^{-1}. (Then summation of the group parameters gave -292.7 kJ·mol^{-1} and -224.9 kJ·mol^{-1} for benzoic acid and naphthalene-2-carboxylic acid, respectively.)

With the above changes in group parameters, the estimation of the ideal-gas enthalpy of formation of DNCA-2M was repeated with the result as follows:

4	C_b–$(C_{bf})(C_b)$(H)	13.82 times 4	55.28
2	C_b–$(C_b)_2$(H)	13.82 times 2	27.64
2	C_{bf}–$(C_{bf})(C_b)_2$	20.10 times 2	40.20
2	C_b–$(C_b)_2$(CO)	17.5 times 2	35.0
2	CO–(C_b)(O)	-136.07 times 2	-272.14
2	O–(CO)(C)	-179.7 times 2	-359.4
2	C–(O)(H)$_3$	-42.29 times 2	-84.58

($\Delta_f H^o_m$ C$_{14}$H$_{12}$O$_4$, g 298.15 K) -558.0 kJ·mol^{-1}.

in excellent agreement with the value determined in this research -555.1 ± 6.1 kJ·mol^{-1} (Table 15).

Table 18. "Assessed" ideal-gas enthalpies of formation for some aliphatic carboxylic acids and the corresponding methyl esters. T = 298.15 K and p° = 101.325 kPa. Values from reference 60. All values in kJ·mol^{-1}.

Acid	$\Delta_f H^o_m$ Acid	$\Delta_f H^o_m$ Methyl ester	Difference [a]
Formic	-378.7 ± 0.6	-355.5 ± 0.8	23.2 ± 1.0
Acetic	-432.8 ± 1.5	-411.9 ± 1.6	20.9 ± 2.2
Pentanoic	-491.9 ± 3.0	-471.2 ± 0.9	20.7 ± 3.1
Hexanoic	-511.9 ± 2.3	-492.6 ± 1.2	19.3 ± 2.6
Heptanoic	-536.2 ± 2.1	-515.9 ± 1.2	20.3 ± 2.4
Octanoic	-554.3 ± 1.5	-533.8 ± 1.3	20.5 ± 2.0
Nonanoic	-577.3 ± 2.1	-553.9 ± 1.9	23.4 ± 2.8
Decanoic	-594.9 ± 2.3	-573.8 ± 1.8	21.1 ± 2.9
Undecanoic	-614.6 ± 1.6	-593.8 ± 1.4	20.8 ± 2.1
Tridecanoic	-660.2 ± 2.5	-635.4 ± 2.7	24.8 ± 3.7
Pentadecanoic	-699.0 ± 4.5	-680.0 ± 2.8	19.0 ± 5.3

[a] Difference = ($\Delta_f H^o_m$ methyl ester $-$ $\Delta_f H^o_m$ acid)

Benzoyl formic acid

As evidenced by both the inclined-piston vapor-pressure measurements and the d.s.c. studies, decomposition started at 429 K in the sample of benzoyl formic acid. Hence, the normal boiling point (532 K) listed in Table 8 is an estimate. Hurd and Ratherink (96) in a study of the decomposition of alpha keto acids note that "Claisen and Bouveault found that benzoyl formic acid yielded benzoic acid and benzaldehyde at 473-523 K." Hurd and Ratherink found CO and CO_2 in the reaction products in addition to benzoic acid and benzaldehyde in studies in the temperature range 523 to 573 K. A search of the literature through 1989 failed to find any thermodynamic property data for benzoyl formic acid.

Using the group parameters given in references 1 and 2 and the new value for the $C_b-(C_b)_2(CO)$ parameter, estimation of the ideal-gas enthalpy of formation of benzoyl formic acid was made as follows:

5	$C_b-(C_b)_2(H)$	13.82 times 5	69.1
1	$C_b-(C_b)_2(CO)$	17.50 times 1	17.5
1	$CO-(CO)(C_b)$	-112.13 times 1	-112.1
1	$CO-(CO)(O)$	-122.59 times 1	-122.6
1	$O-(CO)(H)$	-243.25 times 1	-243.3
($\Delta_f H_m^o$ $C_8H_6O_3$, g 298.15 K)		-391.4 kJ·mol^{-1}	

The result is in good agreement with the value determined in this research -388.7±1.0 kJ·mol^{-1} (Table 15).

Naphthalene-2,6-dicarboxylic acid

As noted above in the Results section of this report, attempts to measure the vapor pressure of naphthalene-2,6-dicarboxylic acid were unsuccessful. At 523 K the vapor pressure was less than the limit of sensitivity of the inclined-piston apparatus (10 Pa). The sample decomposed at 695 K before melting, and hence, no liquid-phase measurements were possible. This relative inertness probably accounts for the complete lack of thermodynamic property data on this acid in the literature (through 1989).

Some thermodynamic property values (heat capacities and the enthalpy of combustion and formation) are reported here in Tables 4, 5, 6, 10, and 15. With the group-additivity parameters given in references 1 and 2 and the parameter revisions made earlier in this report, the ideal-gas enthalpy of formation at 298.15 K was estimated as follows:

4	$C_b-(C_{bf})(C_b)(H)$	13.82 times 4	55.3
2	$C_b-(C_b)_2(H)$	13.82 times 2	27.6
2	$C_{bf}-(C_{bf})(C_b)_2$	20.10 times 2	40.2
2	$C_b-(C_b)_2(CO)$	17.50 times 2	35.0
2	$CO-(C_b)(O)$	-136.07 times 2	-272.1
2	$O-(CO)(H)$	-243.25 times 2	-486.5
($\Delta_f H_m^o$ $C_{12}H_8O_4$, g 298.15 K)		-570.5 kJ·mol^{-1}	

The enthalpy of formation of crystalline naphthalene-2,6-dicarboxylic acid derived in this research (Table 15) is -770.25±0.96 kJ·mol^{-1}. Hence, the enthalpy of sublimation is estimated to be ($\Delta_c^g H_m^o$ $C_{12}H_8O_4$, 298.15 K) = 199.8 kJ·mol^{-1}. This value is considerably higher than the corresponding value for the dimethyl ester ($\Delta_c^g H_m^o$ $C_{14}H_{12}O_4$, 298.15 K) = 146.5±6.0 kJ·mol^{-1} (Table 15). Listings of enthalpies of vaporization in reference 60 also show the aliphatic acids to have larger values than their corresponding methyl esters (data is available through C_{16} where the methyl esters are all liquids at 298.15 K).

Tetraethylsilane

In contrast to the previous few compounds where little or no thermodynamic property measurements were found in the literature search, tetraethylsilane has an overabundance of determined values. Reference 97 is an excellent review of organosilicon thermochemistry through 1988. Particularly in the area of combustion calorimetric determinations, the literature is rife with contradicting values for tetraethylsilane. References 98 through 105 pertain to the combustion calorimetric results which are summarized in Table 19. All of the pre 1971 measurements (98-102) were made in static-bomb calorimeters where incomplete combustion, facilitated by the formation of the fire retardant silicon dioxide, led to erroneous results.

Table 19. Literature values for the enthalpy of formation of tetraethylsilane (T = 298.15 K and $p°$ = 101.325 kPa. All values in kJ·mol^{-1}.)

Reference Number	Year	$\Delta_f H^o_m$
98	1953	-171.5±? [a]
99	1956	-280.0±? [a]
100	1964	-205.0±8.0
101	1966	-205.0±8.0
102	1970	-221.0±21.0
103	1971	-277.8±18.8
104	1986	-335.6±5.6
105	1988	-336.0±4.0
This research	1989	-328.6±5.2

[a] ? = unknown uncertainty interval due to insufficient detail in the reference.

In 1971 Iseard et al. (103) at the University of Sussex, England reported measurements of the energy of combustion of tetraethylsilane using the fluorine additive technique developed by Good et al. (15) in these laboratories, when the standard enthalpy of formation of crystalline silica (silicon dioxide, SiO$_2$) was defined. The University of Sussex group subsequently listed, in a data compilation (106), enthalpies of formation for a number of organosilicon compounds all measured using the fluorine additive technique. A firm foundation for the estimation of the thermochemical properties of organosilicons appeared to have been set. However, group-additivity parameters could not be defined: the results were not internally consistent (97). Subsequent work by Steele (107) on tetramethylsilane (one of the compounds listed in reference 106) gave a significantly different value from that obtained under almost identical calorimetric procedures by the University of Sussex group.

References 104 and 105 appear to report the same research. The enthalpy of formation derived (Table 19) differs substantially from that determined by Iseard et al. (103). The Russian research (104,105) used a "high temperature controlled combustion technique" which produced "highly dispersed hydrated amorphous silicon dioxide" as the sole solid reaction product of the combustion. The authors list a thesis for the method of determination of the enthalpy of formation of the silicon dioxide product of the combustions. A copy of the thesis could not be obtained by the authors of this report. In the absence of the thesis, no definitive comparisons could be made. It would appear that the Russian results are as reliable as the ones obtained at the University of Sussex. The existence of these widely differing values was the reason for the study undertaken in this research.

The energy of combustion measurements reported here were made using the technique developed by Good et al. (15). In that research, hexamethyldisiloxane was also studied. In the research reported here, to test the present procedures, two combustions were performed on the *same* sample of hexamethyldisiloxane as that used in the earlier study. The results gave energies of formation within 0.02 percent of the published value (15).

The derived enthalpy of formation (see Tables 4A, 5, 6, 6A, and 15), $\Delta_f H^o_m$ (C$_8$H$_{20}$Si, l 298.15 K) = -328.6±5.2 kJ·mol^{-1}, is different from all the previous values. However the uncertainty intervals for the result reported here and those listed by Voronkov et al. (104,105) (-335.6±5.6 kJ·mol^{-1} and -336.0±4.0 kJ·mol^{-1}) overlap. The Russian workers also reproduced the enthalpy of formation for tetramethylsilane reported by Steele (107). The agreement appears to not be serendipitous.

The uncertainty in the energy of combustion measurements on tetraethylsilane is higher than those reported in Table 5 for the C,H,N,O compounds studied in this research. This higher uncertainty [of the same order of magnitude as the earlier study on hexamethyldisiloxane (15)] is the penalty for using a combustion reaction in which only a fraction of the evolved energy is produced by the substance of interest.

Ambrose reported (41) a critical temperature of 603 K and a critical pressure of 2602 kPa for tetraethylsilane with no details of the source of the values. In this research, the corresponding experimentally determined critical temperature was 606±1 K and the derived critical pressure was 2400±24 kPa. Literature values for the density of tetraethylsilane in the temperature range 270 K to 430 K (108-110) are compared in Figure 12 with values estimated by extended corresponding states (Equation 15) using the critical properties (this research) listed in Table 9A. The agreement between the various literature values and extended corresponding states is good (±1 percent), and is well within the uncertainty interval assigned to the critical density measured in this research namely, 246±5 kg·m^{-3}.

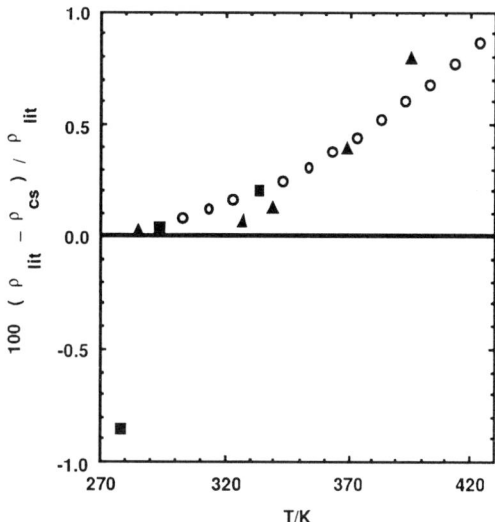

FIGURE 12. Comparison of literature density measurements for tetraethylsilane with those obtained by extended corresponding states. ρ_{cs} = value of the density calculated using equation 15 and the critical properties reported in Table 9A. O; Brostow et al. (108); ▲; Sugden and Wilkins (109); ■; Whitmore et al. (110).

Iseard et al (103) list $(\Delta_l^g H_m^o$ $C_8H_{20}Si$, 298.15 K) = 39.7 kJ·mol^{-1}, quoting reference 110 as the source of the value. However, Whitmore et al. (110) do not give a temperature for their value and state that it was derived from vapor-pressure measurements, only 2 points of which are quoted in the paper although it alludes to many more. Subsequently, Abraham and Irving (111) questioned the value listed by Iseard et al. and using vapor-pressure measurements from the literature (112) derived:

$(\Delta_l^g H_m^o$ $C_8H_{20}Si$, 298.15 K) = 43.3 kJ·mol^{-1}.

The value obtained by Abraham and Irving (111) is in good agreement with the value obtained in this research:

$(\Delta_l^g H_m^o$ $C_8H_{20}Si$, 298.15 K) = 44.62±0.05 kJ·mol^{-1}.

With the results reported in Table 9, the value of 39.7 kJ·mol^{-1} would correspond to a temperature of 369 K and not 298.15 K.

Figure 13 compares values for the vapor pressure of tetraethylsilane found in the literature (110,112) with those calculated using the Cox equation coefficients reported in Table 8. The percentage deviations are within the probable uncertainty intervals of the literature values.

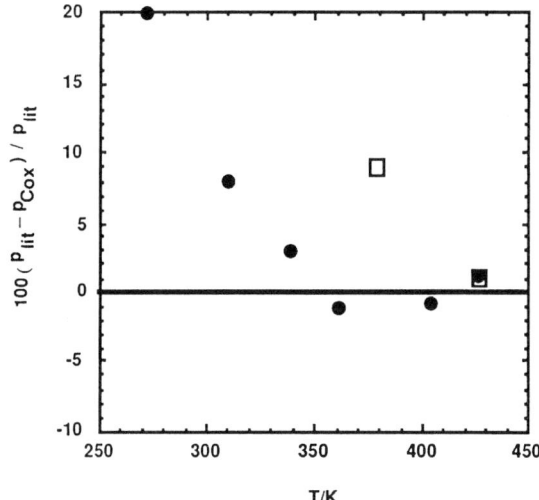

Figure 13. Comparison of literature vapor pressures for tetraethylsilane with values obtained using the Cox equation coefficients reported in Table 8. ●; Stull reference 112; □; Whitmore et al. reference 110.

Group-additivity parameters applicable to tetra-alkyl silanes can be derived using the ideal-gas enthalpies of formation for tetramethylsilane and tetraethylsilane as the anchor points. With the ideal-gas enthalpy of formation at 298.15 K for tetramethylsilane obtained by Steele (107), the Si-(C)$_4$ group can be derived as follows:

4	C-(Si)(H)$_3$	-42.2 times 4	-168.8
		assigned as C-(C)(H)$_3$	
1	Si-(C)$_4$		?

($\Delta_f H_m^o$ C$_4$H$_{12}$Si, g 298.15 K)

-233.2±3.2 kJ·mol^{-1}

and hence, Si-(C)$_4$ equals -64.4 kJ·mol^{-1}. Then the C-(Si)(C)(H)$_2$ group parameter is derived as follows:

4	C-(C)(H)$_3$	-42.2 times 4	-168.8
4	C-(Si)(C)(H)$_2$	? times 4	
1	Si-(C)$_4$		-64.4

($\Delta_f H_m^o$ C$_8$H$_{20}$Si, g 298.15 K)

-284.0±5.2 kJ·mol^{-1},

and hence, C-(Si)(C)(H)$_2$ equals -12.7 kJ·mol^{-1}.

CONCLUSIONS

The group-additivity parameter changes/additions arising from the results reported here can be summarized as follows.

- The C-(C)$_2$(H)$_2$ group-additivity parameter (or the -CH$_2$- increment) is significantly smaller in alkanols than in alkanes.
- The CO-(C$_d$)(N) group-additivity parameter is assigned a value of -117±2 kJ·mol^{-1}.
- The O-(CO)(C) group-additivity parameter is reassigned a value of -179.7 kJ·mol^{-1} from the previous value of -185.5 kJ·mol^{-1}.
- The C$_b$-(C$_b$)$_2$(O) group-additivity parameter is reassigned a value of 17.5 kJ·mol^{-1} from the previous value of 40.6 kJ·mol^{-1}.
- The Si-(C)$_4$ group-additivity parameter is assigned a value of -64.4 kJ·mol^{-1}.
- The C-(Si)(C)(H)$_2$ group-additivity parameter is assigned a value of -12.7 kJ·mol^{-1}.

Results for hexan-1,6-diol could not be reconciled with the listed group-additivity parameters (including the reassigned values). Best estimates overestimate the stability of the compound by 6.7 kJ·mol^{-1}. The discrepancy may be due to 1) misassigned parameter(s), 2) a destabilizing effect within the molecule which is not allowed for in the estimation, or 3) experimental error. Study of a further diol will be required to elucidate the truth.

LITERATURE CITED

1. Benson, S. W. *Thermochemical Kinetics*. 2nd. edition, Wiley, New York, **1976**.
2. Reid, R. C., J. M. Prausnitz, and B. E. Poling *The Properties of Gases and Liquids*. 4th. edition, McGraw-Hill, New York, **1987**.
3. *Pure Appl. Chem.* **1983**, 55, 1101.
4. Cohen, E. R., and B. N. Taylor *J. Phys. Chem. Ref. Data* **1988**, 17, 1795.
5. *Metrologia* **1969**, 5, 35.
6. Good, W. D. *J. Chem. Eng. Data* **1972**, 17. 28.
7. Good, W. D., and N. K. Smith *J. Chem. Eng. Data* **1969**, 14, 102.
8. Good, W. D. *J. Chem. Eng. Data* **1969**, 14, 231.
9. Steele, W. V., D. G. Archer, R. D. Chirico, W. B. Collier, I. A. Hossenlopp, A. Nguyen, N. K. Smith, and B. E. Gammon *J. Chem. Thermodyn.* **1988**, 20, 1233.
10. Good, W. D., D. W. Scott, and G. Waddington *J. Phys. Chem.* **1956**, 60, 1080.
11. Good, W. D., D. R. Douslin, D. W. Scott, A. George, J. L. Lacina, J. P. Dawson, and G. Waddington *J. Phys. Chem.* **1959**, 63, 1133.
12. Guthrie, G. B., D. W. Scott, W. N. Hubbard, C. Katz, J. P. McCullough, M. E. Gross, K. D. Williamson, and G. Waddington *J. Am. Chem. Soc.* **1952**, 74, 4662.
13. Hubbard, W. N., D. W. Scott, and G. Waddington *Experimental Thermochemistry*. Editor Rossini,

F. D. Interscience, New York, **1956**, chap. 5, pp. 75-128.
14. Good, W. D., and R. T. Moore *J. Chem. Eng. Data* **1970**, 15, 150.
15. Good, W. D., J. L. Lacina, B. L. DePrater, and J. P. McCullough *J. Phys. Chem.* **1964**, 68, 579.
16. Good, W. D., D. W. Scott, and G. Waddington *J. Phys. Chem.* **1956**, 60, 1080.
17. Good, W. D., and D. W. Scott *Experimental Thermochemistry.* Vol. 2, Editor Skinner, H. A. Interscience, New York, **1962**, chap. 2, pp. 15-39.
18. Swietoslawski, W. *Ebulliometric Measurements.* Reinhold, New York, **1945**.
19. Osborn A. G., and D. R. Douslin *J. Chem. Eng. Data* **1966**, 11, 502.
20. Chirico, R. D., A. Nguyen, W. V. Steele, M. M. Strube, and C. Tsonopoulos *J. Chem. Eng. Data* **1989**, 34, 149.
21. Antoine, C. *C. R. Acad. Sci.* **1888**, 107, 681.
22. Douslin, D. R., and J. P. McCullough *U.S. Bureau of Mines, Report of Investigation 6149* **1963**, pp. 11.
23. Douslin, D. R., and A. G. Osborn *J. Sci. Instrum.* **1965**, 42, 369.
24. Steele, W. V., D. G. Archer, R. D. Chirico, W. B. Collier, B. E. Gammon, I. A. Hossenlopp, A. Nguyen, and N. K. Smith *The Thermodynamic Properties of Quinoline and Isoquinoline* NIPER-301, March **1988**. Published by DOE Fossil Energy, Bartlesville Project Office. Available from NTIS Order No. DE88001218.
25. Mraw, S. C., and D. F. Naas *J. Chem. Thermodyn.* **1979**, 11, 567.
26. Mraw, S. C., and D. F. Naas *J. Chem. Thermodyn.* **1979**, 11, 585.
27. Mraw, S. C., and D. F. Naas-O'Rourke *J. Chem. Thermodyn.* **1980**, 12, 691.
28. Steele, W. V., R. D. Chirico, S. E. Knipmeyer, and N. K. Smith *High-Temperature Heat-Capacity Measurements and Critical Property Determinations using a Differential Scanning Calorimeter. (Development of Methodology and Application to Pure Organic Compounds)* NIPER-360, December **1988**. Published by DOE Fossil Energy, Bartlesville Project Office. Available from NTIS Order No. DE89000709.
29. Knipmeyer, S. E., D. G. Archer, R. D. Chirico, B. E. Gammon, I. A. Hossenlopp, A. Nguyen, N. K. Smith, W. V. Steele, and M. M. Strube *Fluid Phase Equilibria* **1989**, 52, 185.
30. Rossini, F. D. *Experimental Thermochemistry.* Editor Rossini, F. D. Interscience, New York, **1956**, chap. 14, pp. 297-320.
31. Cox, J. D., D. D. Wagman, and V. A. Medvedev: editors. *CODATA Key Values for Thermodynamics.* Hemisphere: New York **1989**.
32. Kilday, M. V., and E. J. Prosen *J. Res. Nat. Bur. Stand. Sect. A.* **1974**, 77A, 305.
33. Johnson, G. K., P. N. Smith, and W. N. Hubbard *J. Chem. Thermodyn.* **1973**, 5, 793.
34. Scott, D. W., and A. G. Osborn *J. Phys. Chem.* **1979**, 83, 2714.
35. Cox, E. R. *Ind. Eng. Chem.* **1936**, 28, 613.
36. Pitzer, K. S., and R. F. Curl Jr. *J. Am. Chem. Soc.* **1957**, 79, 2369.
37. Hales, J. L., and R. Townsend *J. Chem. Thermodyn.* **1972**, 4, 763.
38. Orbey, H., and J. H. Vera *AIChE J.* **1983**, 29, 107.
39. Steele, W. V., and R. D. Chirico To be submitted to *Ind. Eng. Chem. Res.*
40. Ambrose, D., and R. Townsend *J. Chem. Soc.* **1963**, 3614.
41. Ambrose, D. *Correlations and Estimations of Vapor-Liquid Critical Properties. I. Critical Temperatures of Organic Compounds.* National Physical Laboratory, Teddington, England. *NPL Rep. Chem. 82,* September **1978**, Corrected March **1980**.
Ambrose, D. *Correlations and Estimations of Vapor-Liquid Critical Properties. II. Critical Pressures and Volumes of Organic Compounds.* National Physical Laboratory, Teddington, England. *NPL Rep. Chem. 98,* September **1979**.
42. Chirico, R. D., S. E. Knipmeyer, A. Nguyen, and W. V. Steele *J. Chem. Thermodyn.* **1991**, 23, 431.
43. Goodwin, R. D. *J. Phys. Chem. Ref. Data* **1988**, 17, 1541.

44. Goodwin, R. D. *J. Phys. Chem. Ref. Data* **1989**, 18, 1565.
45. Skinner, H. A., and A. Snelson *Trans. Faraday Soc.* **1960**, 56, 1776.
46. Biddiscombe, D. P., R. R. Collerson, R. Handley, E. F. G. Herrington, J. F. Martin, and C. H. S. Sprake *J. Chem. Soc.* **1963**, 1954.
47. Ambrose, D., and C. H. S. Sprake *J. Chem. Thermodyn.* **1970**, 2, 631.
48. Ambrose, D., and C. H. S. Sprake *J. Chem. Soc. A* **1971**, 1261.
49. Brown, I., W. Fock, and F. Smith *J. Chem. Thermodyn.* **1969**, 1, 273.
50. Andon, R. J., J. E. Connett, J. F. Counsell, E. B. Lees, and J. F. Martin *J. Chem. Soc. A* **1971**, 661.
51. Berkman, N. S., and J. J. McKetta *J. Phys. Chem.* **1962**, 66, 1444.
52. Sellers, P., G. Stridh, and S. Sunner *J. Chem. Eng. Data*, **1978**, 23, 250.
53. Stridh, G., S. Sunner, and Ch. Svensson *J. Chem. Thermodyn.* **1977**, 9, 1005.
54. Cox, J. D. *Trans. Faraday Soc.* **1961**, 57, 1674.
55. Costello, J. M., and S. T. Bowden *Rec. Trav. Chem.* **1958**, 77, 36.
56. Hales, J. L., and J. H. Ellender *J. Chem. Thermodyn.* **1976**, 8, 1177.
57. Thomas, L. H., and R. Meatyard *J. Chem. Soc.* **1963**, 1986.
58. Polák, J., and G. C. Benson *J. Chem. Thermodyn.* **1971**, 3, 235.
59. McCurdy, K. G., and K. J. Laidler *Can. J. Chem.* **1963**, 41, 1867.
60. Pedley, J. B., R. D. Naylor, and S. P. Kirby *Thermochemical Data of Organic Compounds.* 2nd. edition, Chapman and Hall, New York **1986**.
61. Kakiuchi, Y., T. Sakurai and T. Sukuki *J. Phys. Soc. Japan* **1950**, 5, 369.
62. Tanaka, K., T. Seto, A. Wakanable, and T. Hayashida *Bull. Inst. Chem. Res. Kyoto Univ.* **1959**, 37, 281.
63. Frede, E., and D. Precht *Kul. Milchuirtsel. Forchungsber.* **1974**, 26, 325.
64. Precht, D. *Fette Seifen Anstrichm.* **1976**, 78, 145.
65. Precht, D. *Fette Seifen Anstrichm.* **1976**, 78, 189.
66. Yamamoto, T., K. Nozaki, and T. Hara *J. Chem. Phys.* **1990**, 92, 631.
67. Eckert, T., and J. Mueller *Arch. Pharm (Weinheim, Ger.)* **1978**, 311, 31.
68. Mosselman C. *Arch. Pharm (Weinheim, Ger.)* **1981**, 314, 279.
69. Mosselman, C., J. Mourik, and H. Dekker *J. Chem. Thermodyn.* **1974**, 6, 477.
70. Mosselman, C., and H. Dekker *J. Chem. Soc. Faraday Trans. I* **1975**, 71, 417.
71. Davies, M., and B. Kybett *Trans. Faraday Soc.* **1965**, 61, 1608.
72. Spizzichino, C. *J. recherches centre natl. recherche sec. Labs. Bellevue (Paris)* **1956**, 34, 1.
73. Kemme, H. R., and S. I. Kreps *J. Chem. Eng. Data* **1969**, 14, 98.
74. Ambrose, D., and J. Walton *Pure Appl. Chem.* **1989**, 61, 1395.
75. Wagner, W. *Cryogenics* **1973**, 13, 470.
76. Matsao, S., and T. Makita *Int. J. Thermophys.* **1989**, 10, 885.
77. Månsson, M., P. Sellers, G. Stridh, and S. Sunner *J. Chem. Thermodyn.* **1977**, 9, 91.
78. Kramer, C. R. *Z. Phys. Chem. (Leipzig)* **1983**, 264, 82.
79. Vasil'ev, I. A., E. I. Treibsho, A. D. Korkhov, V. M. Petrov, N. G. Orlova, and M. M. Balakina *Inzh.-Fiz. Zh.* **1980**, 39, 1054.
80. Khasanshin, T. S., and T. B. Zykova *Inzh.-Fiz. Zh.* **1989**, 56, 991.
81. Zábransk´y, M., V. R. Rûzicka, Jr., and V. Mayer *J. Phys. Chem. Ref. Data* **1990**, 19, 719.
82. Hoyer, H., and W. Peperle *Z. Electrochem.* **1958**, 62, 61.
83. Gundry, H. A., D. Harrop, A. J. Head, and G. B. Lewis *J. Chem. Thermodyn.* **1969**, 1, 321.
84. Gardner, P. J., and K. S. Hussain *J. Chem. Thermodyn.* **1972**, 4, 819.
85. Knauth, P., and R. Sabbah *J. Chem. Thermodyn.* **1989**, 21, 779.
86. Knauth, P., and R. Sabbah *Can. J. Chem.* **1990**, 68, 731.
87. Watson, K. M. *Ind. Eng. Chem.* **1943**, 35, 398.

88. Byström, K., and M. Månnson *J. Chem. Soc. Perkin Trans. II* **1982**, 565.
89. Lebedeva, N. D., N. M. Gutner, and N. N. Kiseleva *Zh. Org. Khim.* **1976**, 12, 1618.
90. Stull, D. R., E. F. Westrum Jr., and G. C. Sinke *The Chemical Thermodynamics of Organic Compounds*. Wiley & Sons: New York. **1969**.
91. Steele, W. V., R. D. Chirico, A. Nguyen, I. A. Hossenlopp, and N. K. Smith *Determination of Some Pure Compound Ideal-Gas Enthalpies of Formation*. In *Experimental Results from the Design Institute for Physical Property Data: Phase Equilibria and Pure Component Properties-Part II* editors: T. T. Shih and D. K. Jones. *AIChE Symposium Series No. 271 Vol. 85*, **1989**, p.140.
92. Carpenter, E. L., and H. S. Davis *J. Appl. Chem.* **1957**, 7, 671.
93. Freund, M., and K. Fleischer *Ann. Chem.* **1913**, 402, 68.
94. Dozen, Y., S. Fujishoma, and H. Shingu *Thermochim. Acta* **1978**, 25, 209.
95. Colomina, M., M. W. Roux, and C. Turrion *J. Chem. Thermodyn.* **1976**, 8, 869.
96. Hurd, C. D., and H. R. Ratherink *J. Am. Chem. Soc.* **1943**, 56, 1348.
97. Walsh, R. *The Chemistry of Organic Silicon Compounds* Editors; Patai, S., and Z. Rappoport. Wiley, New York, **1988**, chap. 5, pp. 1-21.
98. Tannenbraum, S., S. Kaye, and G. F. Lewenz *J. Am. Chem. Soc.* **1954**, 76, 1027.
99. Lautsch, W. F. *Chem. Tech. (Berlin)* **1958**, 10, 419.
100. Tel'noi, V. I., I. B. Rabinovich, and G. A. Razuvaev *Dokl. Akad. Nauk S.S.S.R.* **1964**, 159, 1106.
101. Tel'noi, V. I., and I. B. Rabinovich *Russ. J. Phys. Chem.* **1966**, 40, 842.
102. Potzinger, P., and F. W. Lampe *J. Phys. Chem.* **1970**, 74, 719.
103. Iseard, B. S., J. B. Pedley, and J. A. Treverton *J. Chem. Soc. A* **1971**, 3095.
104. Voronkov, M. G., V. A. Klyuchnikov, T. F. Danilova, A. N. Korchagina, V. P. Baryshok, and L. M. Landa *Izv. Akad. S.S.S.R. Ser. Khim.* **1986**, 1970.
105. Voronkov, M. G., V. P. Baryshok, V. A. Klyuchnikov, T. F. Danilova, V. I. Pepekin, A. N. Korchagina, and Yu. I. Khudobin *J. Organometallic Chem.* **1988**, 345, 27.
106. Pedley, J. B., and J. Rylance *Sussex-N.P.L. Computer Analysed Thermochemical Data Organic and Organometallic Compounds* University of Sussex, England **1977**.
107. Steele, W. V. *J. Chem. Thermodyn.* **1983**, 15, 595.
108. Brostow, W., T. Grindley, and M. A. Macup *Mater. Chem. Phys.* **1985**, 12, 37.
109. Sugden, S., and H. Wilkins *J. Chem. Soc.* **1931**, 126.
110. Whitmore, F. C., L. H. Sommer, P. A. D. Georgio, W. A. Strong, R. E. Van Strien, D. L. Bailey, H. K. Hall, E. W. Pietruska, and G. T. Kerr *J. Am. Chem. Soc.* **1946**, 68, 475.
111. Abraham, M. H., and R. J. Irving *J. Chem. Thermodyn.* **1979**, 11, 539.
112. Stull, D. R. *Ind. Eng. Chem.* **1947**, 39, 517.

ACKNOWLEDGMENTS

The authors acknowledge Professor E. J. "Pete" Eisenbraun and his research group at Oklahoma State University for purification of six of the eight samples, and the assistance of Jane Thomson in the gas-liquid chromatographic analyses. The authors also acknowledge the helpful discussions with members of the DIPPR Research Project 871 Committee, especially the Chairman Dennis Jones, and Bill Peters of the U.S. Department of Energy Bartlesville Project Office. We gratefully acknowledge the financial support of the U.S. Department of Energy (DOE) and the Design Institute for Physical Property Data (DIPPR). This research was funded within the Supplemental Government Program (SGP) at NIPER as part of the cooperative agreement with the U. S. Department of Energy DE-FC22-83FE60149.

INDEX

A

acentric factor
    tetraethyl silane 116
azeotrope
    azeotrope - single 9,14, 52
    azeotrope - double 3,5

B

Benson groups
    Benson group contributions .... 117 -131
    summary 131
binary mixtures
    binary mixture data 1,6,24,25,32,47

C

critical properties (T,P)
    acetophenone 100
    benzoyl formic acid (and D) 113
    2-butoxyethanol 100
    2-(2-butoxyethoxy) ethanol 100
    ethylene glycol 100
    ethyl-3-ethoxy propionate 100
    hexan -1,6-diol (and D) 113
    1-methoxy 2-propenol acetate 100
    monoethanolamine 100
    naphthalene -2,6-dicarboxylic acid
        dimethyl ester (and D) 113
    2-propoxyethanol 100
    2-(2-propoxyethoxy) ethanol 100
    tetraethylsilane (and D) 113
    valeric acid 100

D

density @ 298.15K
    benzoyl formic acid 104
    ($\pm$) butan-2-ol 104
    hexan-1,6-diol 104
    methacrylamide 104
    naphthalene-2,6-dicarboxylic acid 104
    naphthalene-2,6-dicarboxylic acid
        dimethyl ester 104
    tetradecan-1-ol 104
    tetraethyisliane 104

E

enthalpy of combustion / formation
    benzoyl formic acid
        combustion c; formation c,l,g 109,118
    ($\pm$)-butan -2-ol
        combustion l; formation l,g 109,118
    hexan-1,6-diol
        combustion c; formation c,l,g 109,118
    methacrylamide
        combustion c; formation c,g 109,118
    naphthalene-2,6-dicarboxylic acid
        combustion c; formation c 109,118
    naphthalene-2,6-dicarboxylic acid dimethyl ester
        combustion c; formation c,l,g 109,118
    tetradecan-1-ol
        combustion c; formation c,l,g 109,118
    etraethyl silane
        combustion l, formation l,g 109,118
enthalpy of fusion
    benzoyl formic acid 114
    hexan -1,6-diol 113
    methacrylamide 117
    naphthalene-2,6-dicarboxylic acid
        dimethyl ester 114
    tetradecan-1-ol 113
enthalpy of sublimation
    methacrylamide 117,118
enthalpy of vaporization (T)
    benzoyl formic acid 112
    ($\pm$)-butan-2-ol 112
    hexan -1,6-diol 112
    methacryl amide 117
    naphthalene -2,6-dicarboxylic acid
        dimethyl ester 112
    tetradecan-1-ol 112
    tetraethyl silane 112
equilibria
    liquid-liquid 6
    vapor-liquid 1,6,24,25,32,47

H

heat capacity
    benzoyl formic acid c,l 104,114
    ($\pm$)-butan-2-ol l 104
    hexan-1-6-diol c,l 104,113
    methacrylamide c 104,117
    naphthalene-2,6-dicarboxylic acid c ... 104,114
    naphthalene-2,6-dicarboxylic acid
        dimethyl ester c,l 104,114
    tetradecan-1-ol c,l 104,113
    tetraethyl silane l 104,114,116

S

systems
    acetaldehyde + acetone 1
    acetaldehyde + propylene oxide 1
    acetophenone + propene 6
    aniline + dimethyl ether 47
    ethylbenzylene + n-propyl amine 47
    ethyl cylohexane + sulfolane 32
    1,2 butylene oxide + methyl acetate 1
    1,2 butylene oxide + isobutylene oxide 1
    diethanolamine + water 6
    dimethyl disulfide + methanethiol 6
    dipropylene glycol monomethyl ether
        + 1,2-propandiol 6
    ethyl formate + dimethyl ether 47
    2-methyl tetrahybiofuran
        + tetrahybiofuran 6
    methyl amine + dimethyl ether 47
    n-hexane + benzene 25
    n-octane + sulfolane 32
    o-xylene + sulfolane 32
    phenol + dimethyl ether 47
    phenol + n-propylamine 47

INDEX *continued*

1,2-propanediol + 1-methoxy-2-propanol....6
  propylene oxide + acetophenone.....24
  tetrahydrofuran + 2,3-dihydrofuran......6
  water + 2,5-dihydrofuran...........6

T

thermochemical cycle.................117
triple point
  methacrylamide.....................117

U

UNIFAC
  UNIFAC parameters....................47

V

vapor pressure pure chemicals (experimental)
  acetaldehyde............................3
  acetone..................................3
  acetophenone.......................23,28
  aniline.............................51,65,66
  benzylamine.............................80
  benzoyl formic acid....................111
  (±) butan-2-ol.........................110
  1,2 butylene oxide.......................4
  diacetone alcohol......................80
  2,5 diethanolamine....................21
  2,3-dihydrofuran.......................20
  2,5-dihydrofuran.......................21
  dimethyl disulfide....................23
  dimethyl ether....................54,55,59
  dipropylene glycol monomethyl ether....22
  ethyl benzene......................60,61
  ethyl cyclohexane......................42
  ethyl formate.......................58,59
ethyl-t-butyl ether.........................80
ethyl thioethanol..........................80
hexan-1,6-diol........................110,111
  isobutylene oxide........................4
  methanethiol..........................23
1-methoxy-2-propanol........................22
methacrylamide.............................117
methane sulfonyl chloride...................80
methyacetoacetate..........................80
methyl acetate..............................4
methyl amine............................62,64
2-methyl benzofuran........................80
2-methyl tetrahydrofuran................20,21
N-aminoethyl ethanolamine..................80
naphthalene -2,6-dicarboxylic acid
    dimethyl ether..................111
  n-butylamine...........................80
  n-octane...............................42
n-propyl amine..........................56,57
  o-xylene...............................42
  phenol..............................51,55
  2-phenylpropionaldehyde................80
  1,2-propanediol........................22
propene....................................23
propylene oxide..........................3,28
  sec-butyl acetate......................80
sulfolane..................................42
t-butyl acetate............................80
tetradecan-1-ol...........................110
tetraethyl silane.........................111
tetrahydrofuran........................20,21
water......................................21